**Sustainable gold mining wastew:
by sorption using low-cost 1**

MIKE AGBESI ACHEAMPONG

Thesis committee

Promotor

Prof. Dr P.N.L. Lens
Professor of Environmental Biotechnology
UNESCO-IHE, Delft

Other members

Prof. Dr C.J.N. Buisman, Wageningen University
Prof. P. Le Cloirec, Ecole Nationale Supérieure de Chimie de Rennes, France
Prof. E.D. van Hullebusch, Université Paris-Est Marne-la-Vallée, France
Dr N. Chubar, Glasgow Caledonian University, The United Kingdom

This research was conducted under the auspices of the SENSE Research School for Socio-Economic and Natural Sciences of the Environment

Sustainable gold mining wastewater treatment by sorption using low-cost materials

Thesis
submitted in fulfilment of the requirement of
the Academic Board of Wageningen University and
the Academic Board of the UNESCO-IHE Institute of Water Education
for the degree of doctor
to be defended in public
on Friday, 18 October 2013 at 1:30 p.m.
in Delft, The Netherlands

by

MIKE AGBESI ACHEAMPONG
born in Kumasi, Ghana

Published by:
CRC Press/Balkema
PO Box 11320, 2301 EH Leiden, The Netherlands
e-mail: Pub.NL@taylorandfrancis.com
www.crcpress.com – www.taylorandfrancis.com

ISBN 978-1-138-00165-7 (Taylor & Francis Group)
ISBN 978-94-6173-740-3 (Wageningen University)

Contents

Dedication xiv
Acknowledgements xv
Abstract xvi

1 General introduction 1

   1.1   Background 2
   1.2   Problem Statement 3
   1.3   Research objectives 4
         1.3.1 General objective 4
         1.3.2 Specific objectives 4
   1.4   Research questions 5
   1.5   Outline of the thesis 5
   1.6   References 6

2 Removal of heavy metals from gold mining wastewater - A review 8

   2.1   Introduction 9
         2.1.1 Gold mine wastewater production 9
         2.1.2 Health and environmental risks of heavy metal..............11
         2.1.3 Regulatory discharge limits for wastewater containing heavy metals.
        12
         2.1.4 Scope of this chapter 12
   2.2   Treatment of gold mining wastewater 12
         2.2.1 Physicochemical heavy metal removal methods 13
         2.2.2 Biological heavy metal removal methods.........................15
         2.2.3 Sulphide precipitation combined with biological sulphate reduction..
        15
         2.2.4 Heavy metal sorption and biosorption 18
   2.3   Conclusions 34
   2.4   Acknowledgements 35
   2.5   References 35

3 Assessment of the effluent quality from a gold mining industry in Ghana...45

   3.1   Introduction 46
   3.2   Materials and methods 47
         3.2.1 Study site...47
         3.2.2 Wastewater sampling..49
         3.2.3 Wastewater characterisation 50
   3.3   Results..51
         3.3.1 Variation in process and tailings dam effluents characteristics over
         time 51
         3.3.2 Physical characteristics of process and tailings dam effluents 53
         3.3.3 Chemical characteristics of process and tailings dam effluents.......54
         3.3.4 Principal components analysis 56
   3.4   Discussion 56
         3.4.1 Process and tailings dam effluent characterisation 56
         3.4.2 Pollutants source identification in the process effluent using PCA .58
         3.4.3 Pollutant source identification in the tailings dam effluent using
         PCA 60

3.5   Conclusions 61
3.6   Acknowledgements 61
3.7   References 61

4   Biosorption of Cu(II) onto agricultural materials from tropical regions 65

4.1   Introduction 66
4.2   Materials and methods 67
        4.2.1 Sorption materials 67
        4.2.2 Metal ion solutions 67
        4.2.3 Biosorption experiments 68
        4.2.4 Characterisation of the biosorbents 68
        4.2.5 Analytical techniques 69
        4.2.6 Calculations 69
        4.2.7 Sorption isotherms 70
        4.2.8 Sorption kinetics 72
4.3   Results 73
        4.3.1 Sorption isotherms of Cu(II) 73
        4.3.2 Effect of sorbent particle size 75
        4.3.3 Kinetics of Cu(II) sorption by coconut shell 75
        4.3.4 Physical characterization of the biosorbents 75
        4.3.5 Acid titrations for Point of Zero Charge determination 76
        4.3.6 Fourier transform infrared spectroscopy 77
        4.3.7 Scanning Electron Microscopy and X-ray microanalysis analyses . 79
4.4   Discussion 79
        4.4.1 Physical characterization of the biosorbents 79
        4.4.2 Sorption isotherms 80
        4.4.3 Effect of particle size 81
        4.4.4 Kinetics of Cu(II) sorption by coconut shell 81
        4.4.5 Mechanism of Cu(II) biosorption 81
        4.4.6 Application potential 82
4.5   Conclusions 83
4.6   Acknowledgements 83
4.7   References 84

**5   Cyclic sorption and desorption of Cu(II) onto coconut shell and iron oxide
        coated sand 86**

5.1   Introduction 87
5.2   Materials and Methods 88
        5.2.1 Sorbents 88
        5.2.2 Metal ion and desorption solutions 88
        5.2.3 Sorption experiments 89
        5.2.4 Desorption experiments 89
        5.2.5 Analytical techniques 90
        5.2.6 Calculations 90
5.3   Results 90
        5.3.1 Equilibrium uptake of Cu(II) by CS and IOCS 90
        5.3.2 Screening of desorption solutions 91
        5.3.3 Effect of HCl concentration on Cu(II) desorption 92
        5.3.4 Cyclical Cu(II) sorption and desorption studies 93
5.4   Discussion 94

5.4.1 Effect of desorption cycles on biosorbent capacity 94

5.4.2 Effect of desorption cycles on biosorbent structure.....................95

5.4.3 Effect of desorption solutions 95

5.5   Conclusion 96

5.6   Acknowledgements...97

5.7   References..97

6   Kinetics modelling of Cu(II) biosorption onto coconut shell and *Moringa oleifera* seeds from tropical regions 99

6.1   Introduction...100

6.2   Materials and methods 102

6.2.1 Biosorbents 102

6.2.2 Metal ion solutions...102

6.2.3 Batch biosorption kinetics experiments 102

6.2.4 Analytical techniques...102

6.3   Sorption kinetics models and calculations 102

6.3.1 Sorption kinetics models..102

6.3.2 Calculations...105

6.4   Results...106

6.4.1 Kinetics of Cu(II) biosorption onto coconut shell and *Moringa oleifera* seeds..106

6.4.2 Sorption kinetics modelling 106

6.5   Discussion 112

6.5.1 Kinetics of Cu(II) biosorption onto coconut shell and *Moringa oleifera* seeds..112

6.5.2 Sorption kinetics modelling 112

6.6   Conclusions...113

6.7   Acknowledgement 114

6.8   References..114

7   Removal of Cu(II) by biosorption onto coconut shell in fixed-bed column systems...117

7.1   Introduction...118

7.2   Materials and methods 119

7.2.1 Biosorbent preparation...119

7.2.2 Chemicals...120

7.2.3 Fixed-bed column set-up and experiments 120

7.2.4 Analytical techniques...120

7.2.5 Calculations...121

7.3   Dynamic models 122

7.4   Results...123

7.4.1 Effect of flow rate on Cu(II) biosorption 123

7.4.2 Effect of bed depth on Cu(II) biosorption...................................124

7.4.3 Effect of inlet Cu(II) concentration on Cu(II) biosorption 124

7.4.4 Cu(II) uptake at different column operation parameters................125

7.4.5 Dynamic models 126

7.5   Discussion 128

7.5.1 Cu(II) uptake in the column..128

7.5.2 Effect of operating parameters on breakthrough curves 128

7.5.3 Design of fixed bed sorption columns 129

7.5.4 Stability and reusability of the biosorbent 131
7.6   Conclusions 131
7.7   Acknowledgements 132
7.8   References 132

8   Treatment of gold mining effluent in pilot fixed bed sorption systems 135

8.1   Introduction 136
8.2   Materials and methods 137
          8.2.1 Sorbent preparation 137
          8.2.2 Gold mining effluent 138
          8.2.3 Pilot plant set-up 138
          8.2.4 Experimental design 139
          8.2.5 Analytical techniques 140
          8.2.6 Calculations 141
8.3   Results 141
          8.3.1 Metal removal in CS and IOCS columns 141
          8.3.2 Kinetics of metal uptake and removal efficiency in the treatment
          plant 142
8.4   Discussion 145
          8.4.1 Performance of the two-stage pilot plant treating the gold mining
          effluent 145
          8.4.2 Implication for industrial application 146
8.5   Conclusions 147
8.6   Acknowledgements 148
8.7   References 148

9   General Discussion 150

9.1   Introduction 151
9.2   Gold mining effluent characterisation 151
9.3   Sorption techniques for sustainable gold mining wastewater treatment. 151
9.4   Implication for industrial applications 153
9.5   Further research 154
          9.5.1 Coconut shell (CS) processing and formulation 154
          9.5.2 Effect of chemical treatment on sorption performance of coconut
          shell (CS) for copper 154
          9.5.3 Metal recovery 154
          9.5.4 Disposal of the spent sorbents 155
9.6   Concluding remarks 155
9.7   References 156
Samenvatting 157
List of symbols 160
Greek letters 160
List of publications 161
About the author 163

List of Tables

Table 2-1: Composition of a typical gold mine wastewater from Ghana 11

Table 2-2: Ranking of heavy metals in the CERCLA list of priority chemicals (2005), according to their health and environmental risk, and discharge standards for industrial wastewater in the USA 12

Table 2-3: Advantages and limitations of physicochemical treatments of industrial wastewater 14

Table 2-4: Performance characteristics of some conventional heavy metal removal and recovery technologies (Mark et al., 1995) 15

Table 2-5: Summary of heavy metal removal data by physicochemical treatment methods 16

Table 2-6: Sulphate reduction rates achieved during different reactor runs with various electron donors. 18

Table 2-7: Heavy metal uptake by natural materials through sorption 21

Table 2-8: Data on biosorption of heavy metals using different agricultural and plant biosorbents 23

Table 2-9: Heavy metal uptake by bacterial and algal biomass 24

Table 2-10: Heavy metal uptake by industrial waste through sorption 24

Table 2-11: Frequently used single- and multi-component adsorption models 28

Table 2-12: Data on heavy metal biosorption in fixed-bed column (FBC) and continuous stirred tank reactor (CSTR) 33

Table 3-1: Descriptive statistics of the process effluent characteristics for morning and afternoon taken daily for 75 days. 54

Table 3-2: Descriptive statistics of the tailings dam wastewater characteristics for morning and afternoon taken daily for 75 days 55

Table 3-3: Loading of measured wastewater characteristics (19) on principal components (rotated matrix [a]) for the process effluent (PE) and the Sansu tailings dam effluent (STDW) 59

Table 3-4: Summary of wastewater characteristics from different mineral processing based industrial sources 60

Table 4-1: Coefficients of five different sorption isotherm models for Cu(II) removal by coconut shell, coconut husk, sawdust and *Moringa oleifera* seeds and the coefficient of determination (R^2). 74

Table 4-2: Copper balances over the 24 h sorption experiments with coconut shell, coconut husk, saw dust and *Moringa oleifera* seeds (N=3). 75

Table 4-3: Equilibrium concentrations (C_{eq}) and removal percentages of Cu(II) after sorption onto coconut shell with different particle size ranges. 75

Table 4-4: Physical characteristics of the biosorbents 76

Table 4-5: Normalized element composition of *Moringa oleifera* seeds and coconut shell, obtained with Energy-dispersive X-ray spectroscopy, before (blank) and after sorption experiments. 79

Table 5-1: Physical characterisation of the sorbents used in this study 88

Table 5-2: Uptake of Cu(II) by coconut shell and IOCS. 90

Table 6-1: Physical characteristics of the biosorbents 102

Table 6-2: Summary of the kinetics model parameters for Cu(II) sorption onto the coconut shell and the *Moringa oleifera* seeds 111

Table 6-3: Data on pseudo second-order kinetics rate constant for Cu(II) sorption using different biosorbents 112

Table 7-1:   Physical characteristics of the coconut shell used in this study
              (Acheampong et al., 2011)...120
Table 7-2:   Summary of the models used to evaluate the breakthrough curves in this
              study...123
Table 7-3:   Uptake of Cu(II) at different flow rates..126
Table 7-4:   BDST model parameters for the sorption of Cu(II) onto coconut shell at
              varying bed heights for 10% and 50% saturation...................................127
Table 7-5:   The Thomas, Yoon-Nelson and Clark models parameters for Cu(II)
              biosorption onto coconut shell at different bed heights, flow rate and
              inlet concentration.. 127
Table 8-1:   Characteristics of the gold mining effluent (untreated)........................138
Table 8-2:   Pilot plant operational parameters 139
Table 8-3:   Metal uptake and removal efficiency for the difference runs (bed height
              = 150 cm, flow rate = 150 mL min^{-1}) 143
Table 8-4:   Overall plant performance (columns I and II) for copper, iron and
              arsenic removal..143

List of Figures

Figure 2-1: Schematic diagram of the cyanidation process of gold extraction 10

Figure 2-2: Heavy metals mobilised by man represent a threat as they concentrate in the food chain (modified from Volesky, 2003) 11

Figure 2-3: Schematic representation of physicochemical process to remove metals from solution: Ion-exchange (A), Reversed Osmosis or Ultra filtration (B) and Electro dialysis (C). 13

Figure 2-4: Schematic representation of a hydrogen-fed sulphate-reducing gas-lift bioreactor and settler treating wastewater polluted with heavy metals. ... 17

Figure 2-5: Comparison between the equilibrium concentrations of metal hydroxides and metal sulphides (Huisman et al. 2006) 17

Figure 2-6: Schematic diagram of a two stage process for heavy metal removal by sorption 19

Figure 2-7: A hypothetical sorption gold mining effluent treatment plant 19

Figure 2-8: Schematic flowchart of a sorption process treating gold mining effluent treatment plant (modified from Volesky, 2003) 19

Figure 2-9: A generalised schematic flowchart of biomaterial processing into biosorbent granules with chemical treatment 22

Figure 2-10: Figure 2-9 Coconut biomaterials 25

Figure 2-11: Moringa oleifera biomaterials 25

Figure 2-12: Major binding groups for biosorption (Volesky, 2007) 26

Figure 2-13: Concept of the EMB–SRB system (Modified from Chuichulcherm et al. (2001)). 32

Figure 3-1: Map of the study area showing the Sansu tailings dam of the AngloGold Ashanti mine at Obuasi, Ghana. PE = Process effluent; STPW = Sansu tailings dam wastewater 48

Figure 3-2: Schematic diagram of the tailings dam effluent treatment plant 49

Figure 3-3: Variations in temperature of the process effluent (A) and the wastewater in the Sansu tailings dam (B) measured between June 2010 and September 2010, with Ghana EPA standard (-) 51

Figure 3-4: Variations in pH of the process effluent (A) and the wastewater in the Sansu tailings dam (B) measured between June 2010 and September 2010, with lower (-) and upper (---) limit in Ghana EPA standard 52

Figure 3-5: Variations in copper concentrations of the process effluent (A) and the wastewater in the Sansu tailings pond (B) measured between June 2010 and September 2010, with Ghana EPA Standard (-) 52

Figure 3-6: Variations in arsenic concentrations of the process effluent (A) and the wastewater in the Sansu tailings dam (B) measured between June 2010 and September 2010, with Ghana EPA Standard (-) 52

Figure 3-7: Variations in iron concentrations of the process effluent (A) and the wastewater in the Sansu tailings dam (B) measured between June 2010 and September 2010, with Ghana EPA Standard (-) 53

Figure 3-8: Variations in cyanide concentrations of the process effluent (A) and the wastewater in the Sansu tailings dam (B) measured between June 2010 and September 2010, with Ghana EPA Standard (-) 53

Figure 4-1: Isotherms for Cu(II) sorption onto coconut shell (A), coconut husk (B), saw dust (C) and Moring oleifera seeds (D), obtained by varying the initial Cu(II) concentration. 73

Figure 4-2:   Isotherm for Cu(II) sorption onto coconut shell, obtained by varying the
               sorbent concentration from $10 - 150$ g L^{-1}. 74

Figure 4-3:   Cu(II) sorption onto coconut shell in time (A) and the determination of
               the pseudo first-order (B) and second-order (C) rate constant.. 76

Figure 4-4:   Potentiometric titration curves for coconut shell (A), coconut husk (B),
               saw dust (C) and *Moringa oleifera* seeds (D). 77

Figure 4-5:   FTIR spectra of coconut shell (A), coconut husk (B), saw dust (C) and
               Moring oleifera seeds (D) before (grey) and after Cu(II) sorption
               (black).. 78

Figure 4-6:   Scanning electron micrographs (180 times of magnification) of coconut
               shell (A) and *Moringa oleifera* seeds (B). 79

Figure 5-1:   Sorption and desorption of Cu(II) using coconut shell (CS) and iron
               oxide coated sand (IOCS) as sorbents with HCl, EDTA, acetate, NaOH
               and Ca(NO$_3$)$_2$ as desorption solutions.. 91

Figure 5-2:   Effect of desorbent on particle size of the coconut shell (A, B, C) and
               IOCS (D, E, F); Prior to sorption and desorption application (A, D), after
               cyclical sorption and desorption with 0.05 M HCl (B, E), after cyclical
               sorption and desorption with NaOH (C) and EDTA (F). 92

Figure 5-3:   Sorption (■) and desorption kinetics of Cu(II) from coconut shell using
               0.2 M HCl(◊), 0.1 M HCl (□) and 0.05 M HCl (Δ).. 93

Figure 5-4:   Sorption and desorption cycle of Cu(II) biosorption onto coconut shell
               (■ and □) and IOCS (▲ and Δ).. 93

Figure 5-5:   Desorption cycle of Cu(II) sorption onto coconut shell and IOCS. 94

Figure 6-1:   Cu(II) sorption in time onto coconut shell (A) and *Moringa oleifera*
               seeds (B).. 106

Figure 6-2:   Pseudo first-order plots for Cu(II) biosorption unto coconut shell (A) and
               Moringa oleifera seeds (B). 107

Figure 6-3:   Pseudo second-order plots for Cu(II) biosorption unto coconut shell (A)
               and *Moringa oleifera* seeds (B).. 108

Figure 6-4:   Intraparticle diffusion plots for Cu(II) biosorption unto coconut shell (A)
               and *Moringa oleifera* seeds (B).. 109

Figure 6-5:   Elovich plots for Cu(II) biosorption unto coconut shell (A) and *Moringa
               oleifera* seeds (B). 110

Figure 6-6:   Mass transfer plots for Cu(II) biosorption unto coconut shell (A) and
               Moringa oleifera seeds (B).. 111

Figure 7-1:   Photo and schematic representation of the laboratory-scale fixed-bed
               column setup. 121

Figure 7-2:   Breakthrough curves for Cu(II) biosorption onto coconut shell at
               different flow rates. 124

Figure 7-3:   Breakthrough curves for Cu(II) biosorption onto coconut shell at
               different bed heights.. 125

Figure 7-4:   Breakthrough curves for Cu(II) biosorption onto coconut shell at
               different initial Cu(II) concentrations.. 125

Figure 7-5:   BDST model plots for Cu(II) biosorption onto coconut shell at 10%
               saturation (■) and 50% saturation (▲); at different bed heights. 126

Figure 8-1:   Schematics diagram of the two-stage pilot plant set-up. CS: Coconut
               shell; IOCS: Iron Oxide Coated Sand; EF: Effluent Distributor; P: Pump;
               SP: Sampling Point; V: Valve; GME: Gold Mining Effluent; FM: Flow
               Meter. 140

Figure 8-2:   Schematic representation of the development of dynamic transfer zone in the continuous-flow fixed bed sorption column 140

Figure 8-3:   Amount of copper (Cu) and iron (Fe) remaining in the treated effluent from column I in runs I and 2. 142

Figure 8-4:   Amount of arsenic (As) remaining in the treated effluent from column II.142

Figure 8-5:   Kinetics of copper, iron and arsenic uptake in the treatment plant (columns I + II) for run I (A), run II (B), run III (C) 144

Figure 9-1:   Generalised integrated single stage sorption systems for gold mining effluent treatment using locally available low-cost materials 155

Dedication

This book is dedicated to my lovely wife, Rita Delali Acheampong (Mrs), and our two wonderful children, Manuella Nana Ama Acheampong and Mike Kofi Acheampong Jnr., for their love and support.

Acknowledgements

I would like to express my profound gratitude to my supervisor and promoter, Professor dr. ir. P.N.L. Lens, for his excellent leadership and valuable counsel which enabled me to accomplished this task on schedule. Prof., thank you so much. My sincere thanks go to Dr. Roel JW Meulepas of Westus who was my mentor during the first two years of my PhD studies. Roel, your valuable pieces of advice and unprecedented support during the initial two years of this work was an important catalyst that helped the course of my PhD research during the remaining years of my studies.

I acknowledge collaboration with Professor Kannan Pakshirajan of the Indian Institute of Technology, Indian; Professor Ajit P. Annachhatre of the Asian Institute of Technology, Thailand and Professor Daniel Yeh of the University of South Florida, USA. I owed much appreciation to Ms Joana Pereira and Mr. Anton Dapcic who worked on some aspects of this research during their research fellowship at the UNESCO-IHE Institute for water Education. I would like to extend my gratitude to the UNESCO-IHE laboratory staff for their support during my experimentation period in the laboratory.

To my colleagues and the entire staff of UNESCO-IHE Institute for water Education, I say thank you for the warm friendship and general support. My special thanks to Ms. Jolanda Boots for her kindness and care. It was a valuable experience to work with enthusiastic and dedicated colleagues and staff.

My sencere thanks and appreciation to Dr Henks Lubberding for translating the abstract of my thesis into Dutch (Samenvatting). I also owe Professor Esi Awuah, the Vice Chancellor of the University of Energy and Natural Resources (Ghana) for endorsing my research initiative.

I acknowledge funding from the Netherlands Government under the Netherlands Fellowship Programme (NFP), NUFFIC award (2009-2013), Project Number: 32022513. I also acknowledged funding from the Staff Development and Postgraduate Scholarship Scheme (Kumasi Polytechnic, Ghana) and the UNESCO-IHE Partner Research Fund, UPaRF III research project PRBRAMD (No. 101014). I am very grateful to AngloGold Ashanti (Obuasi mine, Ghana) for providing facilities for the field research work. I am particularly thankful to the manager of the environmental department, Mr. Owusu Yeboah for his guidance and support during the field studies.

My sencere thanks and appreciation go to my wife Rita and children - Manuella and Mike Jnr. for their love and support.

Abstract

Urbanization and industrialization in developing countries has brought about a huge problem of managing both solid and liquid waste and Ghana is no exception. Untreated wastewater from industries and homes eventually end up in rivers and other aquatic ecosystems that are sources of livelihood for humans, flora and fauna. Most of these rivers are used as source of drinking water by rural dwellers without any form of treatment, thus increasing the chances of suffering ill health. In Ghana, the discharge of untreated gold mining wastewater contaminates the aquatic ecosystems with heavy metals such as copper (Cu), threatening ecosystem and human health. The undesirable effects of these pollutants can be avoided by treatment of the mining wastewater prior to discharge. Chapter 2 reviews the relevant parts of technology and biotechnology to remove heavy metals (such as copper, arsenic, lead and zinc) from wastewater. The chapter places special emphasis on gold mining wastewater and the use of low cost materials as sorbent. Various biological as well as physicochemical treatment processes were discussed and compared on the basis of costs, energy requirement, removal efficiency, limitations and advantages. Sorption using natural plant materials, industrial and agricultural wastes has been demonstrated to poses a good potential to replace conventional methods for the removal of heavy metals because of its cost effectiveness, efficiency and local availability of these materials as sorbents. Biosorption is giving prominence as an emerging technology that has been shown to be effective in removing contaminants at very low levels. The parameters affecting (bio)sorption such as initial ion concentration, pH, sorbent dosage, particle size and temperature were discussed. In general, technical applicability, cost-effectiveness and plant simplicity are the key factors in selecting the most suitable treatment method.

An assessment of the wastewater quality is necessary to decide on which contaminant to remove. Chapter 3 presents the quality assessment of the process effluent and the tailings dam effluent of AngloGold-Ashanti Limited, a gold mining company in Ghana. The study showed that the process effluent from the gold extraction plant contains high amounts of suspended solids, and is therefore highly turbid. Arsenic, copper and cyanide were identified as the major pollutants in the process effluent with average concentrations of 10.0, 3.1 and 21.6 mg L^{-1}, respectively. Arsenic, copper, iron and free cyanide (CN^-) concentrations in the process effluent exceeded the Ghana EPA discharge limits. Therefore, it is necessary to treat the process effluent before it can be discharged into the environment. Principal component analysis of the data indicated that the process effluent characteristics were influenced by the gold extraction process as well as the nature of the gold bearing ore processed. The process effluent is fed to the Sansu tailings dam, which removes 99.9% of the Total Suspended Solids and 99.7% of the turbidity; but copper, arsenic and cyanide concentrations were still high. The effluent produced can be classified as inorganic, with a high value of electrical conductivity. Any heavy metal removal technology for this effluent must focus on copper and arsenic. Sorption using agricultural and plant materials or natural sorbents could be a low cost option for treating this effluent.

Sorption is a viable treatment technology for copper-rich gold mine tailings wastewater. In chapter 4, the sorption properties of agricultural materials, namely coconut shell, coconut husk, sawdust and *Moringa oleifera* seeds for Cu(II) were investigated. The Freundlich isotherm model described the Cu(II) removal by coconut husk (R^2 = 0.999) and sawdust (R^2 = 0.993) very well and the Cu(II) removal by

Moringa oleifera seeds (R^2 = 0.960) well. The model only reasonably described the Cu(II) removal by coconut shell (R^2 = 0.932). A maximum Cu(II) uptake of 53.9 mg g^{-1} was achieved using the coconut shell. The sorption of Cu(II) onto coconut shell followed pseudo second-order kinetics (R^2 = 0.997). FTIR spectroscopy indicated the presence of functional groups in the biosorbents, some of which were involved in the sorption process. SEM-EDX analysis suggest an exchange of Mg(II) and K(I) for Cu(II) on *Moringa oleifera* seeds and K(I) for Cu(II) on coconut shell. This study shows that coconut shell can be an important low-cost biosorbent for Cu(II) removal from inorganic wastewater. The results indicate that ion exchange, precipitation and electrostatic forces were involved in the Cu(II) removal by the biosorbents investigated.

For continuous application, the sorbent should be regenerated with an appropriate desorbent, and reused. Chapter 5 presents the findings of the sequential sorption/desorption characteristics of Cu(II) on coconut shell (CS) and iron oxide coated sand (IOCS). In batch assays, CS was found to have a Cu(II) uptake capacity of 0.46 mg g^{-1} and yielded a 93% removal efficiency, while the IOCS had a Cu(II) uptake capacity and removal efficiency of 0.49 mg g^{-1} and 98%, respectively. Desorption experiments indicated that HCl (0.05 M) was an efficient desorbent for the recovery of Cu(II) from CS, with an average desorption efficiency of 96% (sustained for eight sorption and desorption cycles). HCl (0.05 M) did not diminish the CS's ability to sorb copper, even after 8 cycles of sorption and desorption; but completely deteriorated the iron oxide structure of the IOCS within six cycles. This study showed that CS and IOCS are both good sorbents for Cu(II); but cyclical sorption/desorption using 0.05 M HCl is only feasible with CS.

Adsorption kinetic studies are of great significance to evaluate the performance of a given adsorbent and to gain insight into the underlying mechanism. This sorption kinetics of Cu(II) onto coconut shell and *Moringa oleifera* seeds using batch incubations was investigated in chapter 6. In order to understand the mechanisms of the biosorption process and the potential rate controlling steps, kinetic models were used to fit the experimental data. The results indicate that kinetic data were best described by the pseudo second-order model with a correlation coefficient (R^2) of 0.9974 and 0.9958 for the coconut shell and *Moringa oleifera* seeds, respectively. The initial sorption rates obtained for coconut shell and *Moringa oleifera* seeds were 9.6395×10^{-3} and 8.3292×10^{-2} mg g^{-1} min^{-1}, respectively. The values of the mass transfer coefficients obtained for coconut shell ($\beta_1 = 1.2106 \times 10^{-3}$ cm s^{-1}) and *Moringa oleifera* seeds ($\beta_1 = 8.965 \times 10^{-4}$ cm s^{-1}) indicate that the transport of Cu(II) from the bulk liquid to the solid phase was quite fast for both materials investigated. The results showed that intraparticle diffusion controls the rate of sorption in this study; however film diffusion cannot be neglected, especially at the initial stage of sorption.

In chapter 7, the performance of a fixed-bed column packed with coconut shell for the biosorption of Cu(II) ions was evaluated using column breakthrough data at different flow rates, bed-depths and initial Cu(II) concentrations. The Bed Depth Service Time (BDST), Yoon-Nelson, Thomas and Clark models were used to evaluate the characteristic design parameters of the column. The Cu(II) biosorption column had the best performance at 10 mg L^{-1} inlet Cu(II) concentration, 10 mL min^{-1} flow rate and 20 cm bed depth. Under these optimum conditions, the service time to breakthrough and Cu(II) concentration were 58 h and 0.8 mg L^{-1}, respectively, after which the Cu(II) concentration in the effluent exceeded the 1 mg L^{-1} discharge limit set by the Ghana Environmental Protection Agency (EPA). The equilibrium uptake of Cu(II) amounted to 7.25 mg g^{-1}, which is 14.5 times higher than the value obtained in a batch study with the same material for the same initial Cu(II) concentration (10 mg L^{-1}). The BDST model fitted well the experimental data in the 10% and 50% regions of the breakthrough curve. The Yoon-Nelson model predicted well the time required for 50% breakthrough (τ) at all conditions examined. The simulation of the whole breakthrough curve was successful with the Yoon-Nelson model, but the breakthrough curve was best predicted by the Clark model. The design of a fixed bed column for Cu(II) removal from wastewater by biosorption onto coconut shell can be done based on these models.

For industrial applications, the removal of heavy metals from real gold mining effluent (GME) of the AngloGold Ashanti mine (Obuasi, Ghana) in continuous down-flow fixed bed columns using coconut shell and iron oxide-coated sand at a constant flow-rate of 150 mL min^{-1} was studied from December 2011 to June 2012. The treatment plant targeted the removal of copper and arsenic, but other heavy metals (iron, lead and zinc) present in the GME in very low concentrations were also removed. The removal efficiency was greater than 98% in all the cases studied. In one case (run I), a total of 14.8 m^3 of GME was treated in 1,608 h at the time the arsenic breakthrough point occurred in the system. At this point, copper, iron, lead and zinc were completely removed, leaving no traces of the metals in the treated effluent. Copper uptake amounted to 16.11 mg g^{-1}, which is 2.23 times higher that the value obtained in a single ion laboratory column study. Arsenic and iron uptake amounted to 12.68 and 5.46 mg g^{-1}, respectively. The study showed that the down-flow fixed bed treatment configuration is an ideal system for the simultaneous removal of copper and arsenic from low concentration GME, in addition to other heavy metals present in very low concentrations.

Chapter 1

1 General introduction

1.1   Background

Contamination of the environment by heavy metals has become a major concern in recent years (Arief et al., 2008; Zamboulis et al., 2011; Singh et al., 2012). Various industries produce and discharge wastewater containing heavy metals into the environment, posing a serious environmental threat to human health and the ecosystem (Wang and Chen, 2009; Gadd, 2009). Untreated wastewater from industries and homes eventually end up in rivers and other aquatic systems that are a source of livelihood for humans (Saad et al., 2011). Most of these rivers are used as source of drinking water by rural dwellers without any form of treatment, thus increasing the chances of suffering ill health (Ofori, 2010).

The exploration and exploitation of the mineral resources in Ghana has, however, been associated with great environmental and social costs. The extraction of the mineral resources of the country is undertaken by large-scale companies (legal) and by small-scale miners, most of which operate illegally. The activities of the artisanal miners and legal miners contribute to deforestation, loss of biodiversity, land degradation, water pollution, health and environmental impact from the use of chemicals, mostly heavy metals and cyanide in mineral processing. A survey showed that the effects of river pollution occur downstream of a number of mines where there are no effective pollution controls measures.

In 1999, the Environmental Assessment Regulation, Legislative Instrument (L.I.) 1652 was passed in Ghana to ensure that industrial discharges are within acceptable national standards. Startling reports released by the Environmental Protection Agency (EPA) indicate that not even the major industries are discharging acceptable level of wastewater into the aquatic ecosystems, with the mining companies being the worst offenders (Ghana EPA, 2000; 2007; 2010). The EPA of Ghana identified two main reasons for this failure. The first was the lack of adequate human resource capacity of industries to manage their wastewater; and the second was the unavailability of inexpensive wastewater treatment technologies in the country for industries to use. These problems have been captured in Ghana's Growth and Poverty Reduction Strategy document (MEPRC, 2002) for which the Government of Ghana has sought the assistance of her development partners in addressing. As environmental awareness increases, many of the companies are responding to good management practices by signing on to the ISO 14001 and the international cyanide management code.

In recent years, applying biotechnology in controlling and removing heavy metal pollution has received much attention, and gradually became a hot topic in the field of metal pollution control because of its potential application. An alternative process is (bio)sorption, which can be defined as the removal of metal or metalloid species, compounds and particulates from solution by biological materials (Gadd, 1993; Gadd, 2009; Wang and Chen, 2009). The idea of using locally available, low-cost agricultural plant materials for the research, make sorption the preferred method.

The main reason that determines the metal of interest may relate to its chemotoxicity and importance as a pollutant, whether it is a radionuclide or a valuable element (Gadd, 2009). Data obtained from the Obuasi mine of AngloGold Ashanti in Ghana confirmed the presence of heavy metals, with copper, arsenic and cyanide as the major pollutants of interest because of their relatively high concentration and toxicity.

The research will therefore focus on the removal of copper and arsenic from the gold mining wastewater using sorption techniques. The PhD research will involve laboratory investigations as well as field studies on a pilot plant to be constructed at the Obuasi mine of AngloGold Ashanti.

1.2   Problem Statement

Rapid urbanisation and industrialisation in developing countries has brought about a huge problem of managing both solid and liquid waste (Park et al., 2010; Fu and Wang, 2011) and Ghana is no exception. Heavy metal pollution is one of the important environmental problems today (Wang and Chen, 2009). These heavy metals are of special concern due to their toxicity, bioaccumulation tendency and persistency in nature (Garg et al., 2007). In Ghana, environmental groups have for several years accused mining companies of contaminating the environment with heavy metals, arsenic and cyanide. Arsenic and cyanide contamination of streams and rivers in the mining communities of Ghana is a major problem requiring urgent solution. In order to deal effectively with the problem of heavy metal pollution by gold mining activities, the activities of illegal miners must be regulated while the existing legislation on the safe disposal of mining effluent is enforced.

In order to meet stringent regulatory safe discharge standards, it is essential to remove heavy metals from gold mining effluent before they are released into the environment. This means that industries need to develop on-site or in-plant facilities to treat their own effluents and minimise contaminants concentration to acceptable limits prior to discharge (Banat et al., 2002). The conventional methods for removing heavy metal ions from wastewater include chemical precipitation, coagulation, filtration, ion-exchange, reverse osmosis, membrane filtration, evaporation recovery and electrochemical technologies (Anurag et al., 2007; Meena et al., 2008; Akar and Tunali, 2005; Ahluwalia and Goyal, 2007). However, conventional treatment technologies like precipitation and coagulation become less efficient and more expensive when situations involving high volume and low metal concentrations are encountered, and the application of membrane processes and activated carbon are restricted due to high costs (Anurag et al., 2007). Biosorption may provide an alternative low cost answer to these problems (Apiratikul and Pavasant, 2008).

There is, thus, a need to develop low-cost technologies based on locally available and cheap sorption materials. Coconut husk and shells are agricultural plant materials and are generally disposed off into the environment. These materials are available in abundant quantities worldwide in wet tropical areas (Van Dam et al., 2004) such as Ghana.
In Ghana, the indiscriminate disposal of the waste (especially in the big cities like Accra and Kumasi) is a major environmental concern for the city authorities with regard to the management of those solid wastes. It is anticipated that this research would help solve, indirectly, the environmental problems caused by these waste materials if the coconut husk and shell are used as sorbent for the removal of different contaminants from gold mining wastewater.

The potential of coconut husk, coconut shell, *Moringa oleifera* seeds and sawdust from a local tree (*Triplochiton scleroxylon*) as low cost sorbent will be studied, with greater attention paid to the understanding of the sorption mechanism of the metals, chemical speciation and binding sites involved in the sorption process; as these areas have not been well exploited and understood. Another adsorbent of interest is Iron Oxide Coated Sand (IOCS), which is a by-product of water treatment plants and as such it is widely available and inexpensive. In this regard, the research aims at developing cost effective and appropriate wastewater treatment technology for the removal of copper and arsenic from gold mining effluent.

1.3   Research objectives

1.3.1   General objective
The research was aimed at the design, construct and operation a pilot plant for the removal of arsenic, copper and other heavy metals from gold mining wastewater using low-cost sorbents. An evaluation of the overall performance of the treatment system in order to establish its potential for industrial application was necessary. It was expected that arsenic and copper would be successfully removed from the effluent by the studied materials through the application of sorption technique. The research considered the down-flow fixed-bed treatment configuration for the simultaneous removal of copper and arsenic from the low concentration gold mining effluent, in addition to other heavy metals present in very low concentrations.

1.3.2   Specific objectives
The specific objectives of this research were to:

1. Quantitatively assess the quality of gold mining wastewater of a selected gold mine in Ghana with respect to heavy metals, cyanide, sulphate ions, carbonate ions, COD, BOD and other relevant inorganic compounds.
2. Screen four selected sorbents (Coconut Husk, Coconut Shell, *Moringa oleifera* Seeds (MOS), and Sawdust (SD) from *Triplochiton scleroxylon* wood) to determine their sorption properties (capacity and kinetics) in order to select the most promising one with sufficiently high sorption capacity for this research.
3. Optimize the conditions for maximum sorption by studying the effects of process parameters such as concentration of the influent, contact time, adsorbent dosage, particle size, etc. on the removal of the copper and arsenic by the selected sorption material.
4. Study the removal mechanism of heavy metals by the selected sorption material. Kinetics of the process will be studied in details using existing kinetics models.
5. Study the removal of the heavy metals by the selected sorption materials in a laboratory column experiment. The potential of iron-oxide-coated sand (IOCS) as low-cost adsorbents will also be evaluated. The breakthrough curve of the system will be established.
6. Design, construct and operate a pilot plant at the Obuasi mine of AngloGold Ashanti Limited in Ghana based on the results of the laboratory scale investigations and literature. Evaluate the overall performance of the treatment system in order to establish its potential for industrial application.

1.4   Research questions

In order to achieve the objectives of the research, it was important to provide answers to the following questions:

1. What are the pollutants of importance in the gold mining effluent
2. Which of the proposed sorption materials have the adequate properties (capacity and kinetics) for use in this research?
3. What is the effect of process parameters (metal concentrations, amount of sorbent and contact time) on sorption?
4. Which functional groups are involved in the binding of metal ions?
5. What is the mechanism of metal removal?
6. What are the optimum operation parameters for efficient removal of heavy metals?

1.5   Outline of the thesis

This thesis has been divided into nine chapters. The current chapter gives a background to the dissertation including the problems associated with the use of conventional techniques for wastewater treatment. Furthermore, it gives the objectives to be addressed; the relevance and the expected output of the study.

Chapter 2 reviews the relevant parts of technology and biotechnology to remove heavy metals (such as copper, arsenic, lead and zinc) from wastewater. The chapter places special emphasis on gold mining wastewater and the use of low cost materials as sorbent.

Chapter 3 presents the quality assessment of the process effluent and the tailings dam effluent of AngloGold Ashanti Limited, a gold mining company in Ghana. The effluent produced can be classified as inorganic, with a high load of non-biodegradable compounds. Any heavy metal removal technology for this effluent must focus on copper and arsenic.

In Chapter 4, the sorption properties of agricultural materials, namely coconut shell, coconut husk, sawdust and *Moringa oleifera* seeds for Cu(II) were investigated.

Chapter 5 presents the findings of the sequential sorption/desorption characteristics of Cu(II) on coconut shell (CS) and iron oxide coated sand (IOCS).

The sorption kinetics of Cu(II) onto coconut shell and *Moringa oleifera* seeds using batch incubations was investigated in Chapter 6. In order to understand the mechanisms of the biosorption process and the potential rate controlling steps, kinetic models were used to fit the experimental data.

In Chapter 7, the performance of a fixed-bed column packed with coconut shell for the sorption of Cu(II) ions was evaluated using column breakthrough data at different flow rates, bed-depths and initial Cu(II) concentrations. The Bed Depth Service Time (BDST), Yoon-Nelson, Thomas and Clark models were used to evaluate the characteristic design parameters of the column.

Chapter 8 presents the removal of heavy metals from real gold mining effluent (GME) of the AngloGold Ashanti mine (Obuasi, Ghana) in continuous down-flow fixed bed pilot columns using coconut shell and iron oxide-coated sand as sorbents. The pilot treatment plant targeted the removal of copper and arsenic, but other heavy metals (iron, lead and zinc) present in the GME in very low concentrations were also removed.

Finally, Chapter 9 summarises the results of the study and gives recommendation for practice and future research.

1.6   References

Akar, T., Tunali, S., 2005. Biosorption performance of *Botrytis cinerea* fungal by-products for removal of Cd(II) and Cu(II) ions from aqueous solutions. Minerals Engineering 18, 1099-1109.

Ahluwalia, S.S., Goyal, D., 2007. Microbial and plant derived biomass for removal of heavy metals from wastewater. Bioresource Technology 98, 2243-2257.

Anurag Pandey, D.B.A.S., Lalitagauri, R., 2007. Potential of Agarose for Biosorption of Cu(II) In Aqueous System. Science Publications.

Apiratikul, R., Pavasant, P., 2008. Batch and column studies of biosorption of heavy metals by Caulerpa lentillifera. Bioresource Technology 99, 2766-2777.

Arief, V.O., Trilestari, K., Sunarso, J., Indraswati, N., Ismadji, S., 2008. Recent Progress on Biosorption of Heavy Metals from Liquids Using Low Cost Biosorbents: Characterization, Biosorption Parameters and Mechanism Studies. CLEAN – Soil, Air, Water 36, 937-962.

Atkinson, B.W., Bux, F., Kasan, H.C., 1998. Considerations for application of biosorption technology to remediate metal-contaminated industrial effluents. *Water SA* **24**, 129-35.

Banat, F., Al-Asheh, S., Mohai, F., 2002. Multi-Metal Sorption By Spent Animal Bones. Separation Science and Technology 37, 311-327.

Fu, F., Wang, Q., 2011. Removal of heavy metal ions from wastewaters: A review. Journal of Environmental Management 92, 407-418.

Gadd, G.M., 2009. Biosorption: critical review of scientific rationale, environmental importance and significance for pollution treatment. Journal of Chemical Technology & Biotechnology 84, 13-28.

Gadd, G.M., White, C., 1993. Microbial treatment of metal pollution - a working biotechnology? Trends in Biotechnology 11, 353-359.

Ghana EPA, 2000. Annual report: Industrial effluent monitoring. Environmental Protection Agency, Accra, Ghana.

Ghana EPA, 2007. Annual report: Industrial effluent monitoring. Environmental Protection Agency, Accra, Ghana.

Ghana EPA, 2010. Environmental performance rating and disclosure: report on the performance of mining and manufacturing companies. Environmental Protection Agency, Accra, Ghana.

Meena, A.K., Kadirvelu, K., Mishra, G.K., Rajagopal, C., Nagar, P.N., 2008. Adsorptive removal of heavy metals from aqueous solution by treated sawdust (Acacia arabica). Journal of Hazardous Materials 150, 604-611.

MEPRC, 2002. *Growth and Poverty Reduction Strategy (GPRS II)*. Ministry of Economic Planning and Regional Cooperation, Government of Ghana, Accra.

Ofori, J., 2010. Microbial pollution studies in selected inland water bodies: the Densu basin. *CSIR Water Research Institute Annual Report, 2010*. Council for Scientific an Industrial Research, Accra.

Saad, D.M., Cukrowska, E.M., Tutu, H., 2011. Development and application of cross-linked polyethylenimine for trace metal and metalloid removal from mining and industrial wastewaters. Toxicological & Environmental Chemistry 93, 914-924.

Singh, A., Kumar, D., Gaur, J.P., 2012. Continuous metal removal from solution and industrial effluents using Spirogyra biomass-packed column reactor. Water Research 46, 779-788.

van Dam, J.E.G., van den Oever, M.J.A., Keijsers Edwin, R.P., 2004. Production process for high density high performance binderless boards from whole coconut husk. Industrial Crops and Products 20, 97-101.

Wang, J., Chen, C., 2009. Biosorbents for heavy metals removal and their future. Biotechnology Advances 27, 195-226.

Zamboulis, D., Peleka, E.N., Lazaridis, N.K., Matis, K.A., 2011. Metal ion separation and recovery from environmental sources using various flotation and sorption techniques. Journal of Chemical Technology & Biotechnology 86, 335-344.

Chapter 2

2 Removal of heavy metals from gold mining wastewater - A review

The main part of this chapter was published as:

Acheampong, M.A., Meulepas, R.J., Lens, P.N., 2010. Removal of heavy metals and cyanide from gold mine wastewater. Journal of Chemical Technology & Biotechnology 85, 590-613.

Abstract

This chapter reviews the relevant parts of technology and biotechnology to remove heavy metals (such as copper, arsenic, lead and zinc) from contaminated wastewater. The chaper places special emphasis on gold mine wastewater and the use of low cost materials as sorbent. Various biological as well as physicochemical treatment processes were discussed and compared on the basis of costs, energy requirement, removal efficiency, limitations and advantages. Sorption using natural plant materials, industrial and agricultural wastes has been demonstrated to poses a good potential to replace conventional methods for the removal of heavy metals because of its cost effectiveness, efficiency and local availability of these materials as biosorbent. The parameters affecting sorption such as initial ion concentration, pH, sorbent dosage, particle size and temperature were discussed. The overall treatment cost of metal contaminated wastewater depends on the process employed and the local conditions. In general, technical applicability, cost-effectiveness and plant simplicity are the key factors in selecting the most suitable treatment method.

Keywords: removal, sorption, heavy metal, cyanide, gold mine wastewater

2.1 Introduction

The term heavy metal refers to metallic elements with relatively high densities that are toxic at low concentrations. Heavy metals have atomic weights between 63.5 and 200.6, and a specific gravity higher than 5.0 (Srivastava and Majmder, 2008). Heavy metals are classified into three main groups as toxic metals (such as Hg, Cr, Pb, Zn, Cu, Ni, Cd, As, Co, Sn etc.), precious metals (such as Pd, Pt, Ag, Au, Ru etc) and radionuclides (such as U, Th, Ra, Am etc.) (Wang and Chen, 2009). Contamination of wastewaters by toxic metal ions is a worldwide environmental problem. The main sources of pollution are mining and electroplating industries discharging a variety of toxic metals such as Pb, Cu, Ni, Zn, As, and Cd ions into soils and water bodies (Ahmady-Asbchin et al., 2008). Gold mine wastewater generally contains heavy metal pollutants (such as Cu, As, Fe, Zn, Pb etc) and cyanide at elevated concentrations. The toxicity of both metals and cyanide pollution is long lasting as these pollutants are non-biodegradable (Gipta et al., 2001; Shukla and Pai, 2005).

2.1.1 Gold mine wastewater production

Gold is generally extracted from ores or concentrates by the Alkaline Cyanidation (Elsner) Process (Figure 2-1). The gold-bearing ore is crushed and ground to approximately 100 microns. Next, it is transported to a leaching plant where lime, cyanide and oxygen are added to the ground and slurred ore. The lime raises the pH, while the oxygen and cyanide oxidize and complex the gold (see equation 2-1):

$$4Au + 8NaCN + O_2 + 2H_2O \rightarrow 4\ Na[\ Au(CN)_2\ ] + 4NaOH \qquad (2\text{-}1)$$

The cyanide solution thus dissolves the gold from the crushed ore. Next, the gold-bearing solution is collected. Finally, the gold is precipitated out of the solution.

The common processes for recovery of the dissolved gold from solution are carbon-in-pulp, the merrill-crowe process, electrowinning and resin-in-pulp (Teeter and Houck, unpublished). In the carbon-in-pulp (CIP) technique, the gold cyanide complex is adsorbed onto activated carbon (Figure 2-1) until it comes to equilibrium with the gold in solution. Because the carbon particles are much larger than the ore particles, the coarse carbon can be separated from the slurry by screening using a wire mesh. The gold loaded carbon is then removed and washed before undergoing "elution" or desorption of gold cyanide at high temperature and pH. The rich eluate solution that emerges from the elution process is passed through electrowining cells where gold and other metals are precipitated onto the cathodes. Smelting of the cathodes material further refines the gold and produces gold ingots suitable for transport to a refinery (Ripley et al., 1996). Mercury amalgamation and gravity concentration are the other processes for obtaining gold concentrate from gold ore.

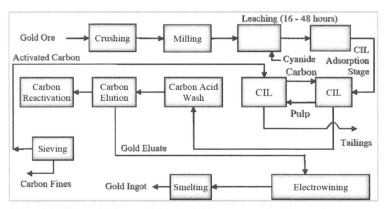

Figure 2-1: Schematic diagram of the cyanidation process of gold extraction

The tailings (Figure 2-1), contaminated with metal and cyanide ions, are usually stored in tailings ponds; with the potential for ground water contamination and high risk of failure, which can lead to spillage of the toxic metals and cyanide bearing solution into the environment. The types of heavy metals present depend on the nature of the gold ore. Acid mine drainage is another type of mine effluent, which is produced when sulphide ores are exposed to the atmosphere as a result of mining and milling processes where oxidation reactions are initiated. Mining increases the exposed surface area of sulphur-bearing rocks allowing for excess acid generation beyond the natural buffering capabilities found in host rock and water resources. Collectively, the generation of acidity from sulphide weathering is termed Acid Mine Drainage (AMD).

Table 2-1: Composition of a typical gold mine wastewater from Ghana

Parameter	Typical gold mining wastewater	wastewater standards (Ghana EPA, 2007)
pH	7.40	6-9
Conductivity (µS/cm)	5600	750
TDS (mg L^{-1})	2900	50
TSS (mg L^{-1})	22	50
Temperature (^{0}C)	31.3	< 40
Cyanide (mg L^{-1})	9	0.2
As (mg L^{-1})	7.350	0.2
Fe (mg L^{-1})	0.114	2.0
Pb (mg L^{-1})	0.140	0.1
Cu (mg L^{-1})	5.063	1.0
Zn (mg L^{-1})	0.042	2.0

Concentrations of common elements such as Cu, Zn, Al, Fe, As and Mn all dramatically increase in waters with low pH (Jennings et al., 2008). Table 2-1 shows the characteristics of a typical goldmine wastewater.

2.1.2 Health and environmental risks of heavy metal

Heavy metal pollution is one of the important environmental problems today (Wang and Chen, 2009). These heavy metals are of special concern due to their toxicity, bioaccumulation tendency and persistency in nature (Srivastava and Majumder, 2008; Randall et al., 1979; Garg et al., 2007). The removal of heavy metals from water and wastewater is important in terms of protection of public health and environment due to their accumulation in living tissues through the food chain (Figure 2-2) as a non-biodegradable pollutant (Sari et al., 2007; Gündogan et al., 2004). Heavy metals (such as lead, copper and arsenic) are toxic to aquatic flora and fauna even in relatively low concentrations (Mohan et al., 2007).

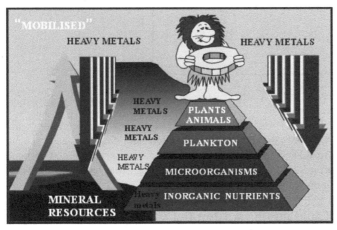

Figure 2-2: Heavy metals mobilised by man represent a threat as they concentrate in the food chain (modified from Volesky, 2003)

The excessive intake of copper by man leads to severe mucosal irritation, widespread capillary damage, hepatic and renal damage, central nervous problems followed by depression, gastrointestinal irritation and possible necrotic changes in the liver and kidney (Sengil et al., 2008; Han et al., 2009). On the other hand, arsenic dissolved in water is acutely toxic and leads to a number of health problems; including disturbances in the cardiovascular and nervous system functions and eventually death (Boddu et al., 2008). Other heavy metals (such as Hg, Cd, Se, Pb, Ni, Zn, etc.) produce similar health effects when injected in significant quantities. Clearly, heavy metal pollution of the environment is of paramount concern due to their health risk to humans and threat to the ecosystem.

2.1.3   Regulatory discharge limits for wastewater containing heavy metals

Gold mining operations result in contamination of soils and water with tailings that release toxic metals such as Cu, As, Pb, Mo, Fe, Ni and Zn (Schneegurt et al., 2001). Various regulatory bodies have set the maximum prescribed limit for discharge of toxic heavy metals into the aquatic ecosystem. Table 2-2 shows the maximum contaminants level (MCL) values as per US EPA and the position in the Comprehensive Compliance Environmental Response, Compensation and Liability Act (CERCLA), 2005 list of priority chemical of some of the toxic heavy metals (Srivastava and Majumder, 2008; USEPA, 2005). Nevertheless, the metal ions are discharged into the water bodies at much higher concentrations than the prescribed limit by industrial activities such as gold mining, thus leading to health hazards and environmental degradation (Sud et al., 2008).

Table 2-2: Ranking of heavy metals in the CERCLA list of priority chemicals (2005), according to their health and environmental risk, and discharge standards for industrial wastewater in the USA

Heavy metal	Rank	Standards (USA) (mg L^{-1})
Arsenic (As)	01	0.01
Lead (Pb)	02	0.015
Mercury (Hg)	03	0.002
Cadmium (Cd)	08	0.005
Chromium (Cr (VI))	18	0.01
Zinc (Zn)	74	5.0
Manganese (Mn)	115	0.05
Copper (Cu)	133	1.3
Selenium (Se)	147	0.05
Silver (Ag)	213	0.05
Antimony (Sb)	222	0.006
Iron (Fe)	-	0.3

2.1.4   Scope of this chapter

The purpose of this chapter is to review the relevant parts of technology and biotechnology to remove heavy metals from wastewater. The chapter places special emphasis on gold mining wastewater and the use of low cost biomaterial s as sorbent. Biosorption is giving prominence as an emerging technology that has been shown to be effective in removing contaminants at very low levels.

2.2   Treatment of gold mining wastewater

The undesirable effects of heavy metals pollution can be avoided by treatment of the wastewater prior to discharge (Monsser et al., 2002). In view of their toxicity and in order to meet regulatory safe discharge standards, it is essential to remove heavy metals from gold mining wastewater before it is released into the environment. This helps to protect the environment and guarantee quality public health. The available treatment methods are reviewed below.

2.2.1   Physicochemical heavy metal removal methods

The conventional methods for removing heavy metal ions from wastewater include chemical precipitation, coagulation-flocculation, floatation, filtration, ion-exchange, reverse osmosis, membrane-filtration, evaporation recovery and electrochemical technologies (Anurag et al., 2007; Meena et al., 2008; Rich and Cherry, 1987; Akar and Tunali, 2005; Bailey et al., 1999; Ahluwalia and Goyal, 2007 and Esposito et al., 2001). Mavrov et al (1995) used a hybrid process of electrocoagulation and membrane filtration (MF) to remove heavy metals from industrial wastewater. Although these treatment methods can be employed to remove heavy metals from contaminated wastewater, they have their inherent advantages and limitations in application (see Table 2-3). Figure 2-3 shows the schematic representation of physicochemical process to remove metals from solution by ion-exchange, reverse osmosis, Ultrafiltration and electrodialysis.

Precipitation is most applicable among these techniques and considered to be the most economical (Monsser and Adhoum, 2002). However; this technique produces a large amount of sludge precipitate that requires further treatment. Reverse osmosis and ion-exchange can effectively reduce the metal ions, but their use is limited due to a number of disadvantages such as high materials and operational cost, in addition to the limited pH range for the ion-exchange resin (Anurag et al., 2007; Kapoor and Viraraghavan). Table 2-4 gives performance characteristics of some conventional heavy metal removal and recovery technologies while Table 2-5 summarises research work on heavy metal removal using physicochemical techniques.

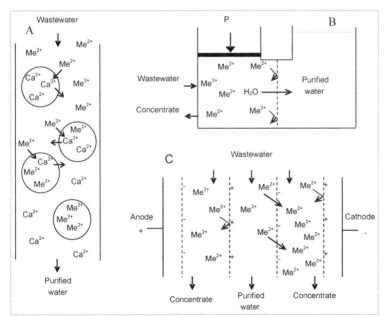

Figure 2-3: Schematic representation of physicochemical process to remove metals from solution: Ion-exchange (A), Reversed Osmosis or Ultra filtration (B) and Electro dialysis (C).

Table 2-3: Advantages and limitations of physicochemical treatments of industrial wastewater

Type of treatment	Target of removal	Advantages	Disadvantages	References
Reverse osmosis	Organic and inorganic Compounds	High rejection rate, able to withstand high temperature	High energy consumption due to high pressure required (20–100 bar), susceptible to membrane fouling	Potts et al. (1981); Kurniawan et al. (2006)
Electrodialysis	Heavy metals	Suitable for metal concentration less than 20 mg L^{-1}	Formation of metal hydroxide, high energy cost, cannot treat a metal concentration higher than 1000 mg L^{-1}	Bruggen and Vandecasteele (2002); Ahluwalia and Goyal (2007)
Ultrafiltration	High molecular weight compounds (1000–10000 Da)	Smaller space requirement	High operational cost, prone to membrane fouling, generation of sludge that has to be disposed off	Vigneswaran et al. (2004)
Ion exchange	Dissolved compounds, cations/anions	No sludge generation, less time consuming	Not all ion exchange resins are suitable for metal removal, high capital cost	Vigneswaran et al. (2004); Ahluwalia and Goyal (2007); Kurniawan et al. (2006)
Chemical precipitation	Heavy metals, divalent metals	Low capital cost, simple operation	Sludge generation, extra operationalcost for sludge disposal	Ahluwalia and Goyal. (2007); Bose et al. (2002); Wingenfelder et al. (2005)
Coagulation–flocculation	Heavy metals and suspended Solids	Shorter time to settle out suspended solids, improved sludge settling	Sludge production, extra operationalcost for sludge disposal	Shammas (2004); Semerjian and Ayoub (2003); Ayoub et al. (2001)
Dissolved air flotation	Heavy metals and suspended Solids	Low cost, shorter hydraulic retention time	Subsequent treatments are required to improve the removal efficiency of heavy metal	Lazaridis et al. (2001)
Nanofiltration	Sulphate salts and hardness ions such as Ca(II) and Mg(II)	Lower pressure than RO(7–30 bar)	Costly, prone to membrane fouling	Ahn et al. (1999)
Electrochemical precipitation	Heavy metals	Can work under both acidic and basic conditions, can treat effluent with a metal concentration higher than 2000 mg L^{-1}	High capital and operational costs	Subbaiah et al. (2002)
Membrane electrolysis	Metal impurities	Can treat wastewater with metal concentration of less than 10 mg/l or higher than 2000 mg L^{-1}	High energy consumption	Kurniawan et at. (2006)

Table 2-4: Performance characteristics of some conventional heavy metal removal and recovery technologies (Mark et al., 1995)

Technology	Performance characteristics				
	pH change	Metal selectivity	Influence of suspended solids	Tolerance to organic molecules	Metal working level (mg L^{-1})
Adsorption (e.g. GAC*)	Limited tolerance	Moderate	Fouled	Can be poisoned	<10
Electrochemical	Tolerant	Moderate	Can be engineered to tolerate	Can be accommodated	>10
Ion exchange	Limited tolerance	Some selectivity (e.g. chelating resin)	Fouled	Can be poisoned	<100
Precipitation as hydroxide	Tolerant	Non-selective	Tolerant	Tolerant	>10
Solvent extraction	Some tolerant systems	Metal-selective extractants available	Fouled	Intolerant	>100

*Granular Activated Carbon

2.2.2 Biological heavy metal removal methods

In recent years, research attention has focused on biological methods for the treatment of metal bearing effluents. There are three principal advantages of biological technologies for the removal of pollutants. First, biological processes can be carried out *in situ* at the contaminated site. Second, bioprocess technologies are usually environmental benign (no secondary pollution). Third, they are cost effective (Volesky and Holan, 1995; Chandra et al., 1998; Vijayaraghavan and Yun, 2008).

2.2.3 Sulphide precipitation combined with biological sulphate reduction

Sulphate reduction is mediated by a diverse group of prokaryotes that can gain energy for growth and maintenance from the reduction of sulphate to sulphide (Castro et al., 2000). For sulphate reduction an electron donor is required. Sulphate reducers are able to use a wide range of electron donors, including hydrogen, formate, methanol, ethanol, fatty acids and sugars (Muyzer and Stams, 2008). Sulphate reduction only occurs when electron acceptors with a higher redox potential (e.g. oxygen and nitrate) are absent. In nature, such conditions are found in marine and fresh-water sediments (Postgate, 1984).

Sulphate reduction in anaerobic bioreactors treating organic wastes has long been regarded as an unwanted side process due to the loss of electron donor and inhibition of the methanogenic process by sulphide (Colleran et al., 1995). However, currently biological sulphate reduction is an established biotechnological process for the treatment of inorganic wastewaters containing oxidized sulphur compounds and heavy metals (Weijma et al., 2002, Lens et al., 2002). The wastewater and an electron donor are fed into a reactor containing immobilized sulphate reducing microorganisms. In this bioreactor, sulphate is reduced and the dissolved metals precipitate with the biologically produced sulphide (Figure 2-4). The formed insoluble metal sulphides can subsequently be separated from the water phase in a settler. Figure 2-5 compares the equilibrium concentrations of metal hydroxides and metal sulphides.

Table 2-5: Summary of heavy metal removal data by physicochemical treatment methods

Treatment method	Metal	Initial metal conc. (mg L^{-1})	pH	Removal efficiency (%)	Power consumption (kWh m^{-3})	References
Chemical precipitation	Zn(II)	450	11.0	99.77	NA	Cherernnyavak (1999)
	Cd(II)	150	11.0	99.76	NA	
	Mn(II)	1085	11.0	99.30	NA	
	Cu(II)	16	9.5	80	NA	Tünay and Kabdalsi (1994)
Electrochemicalcoagulation/Membrane Filtration	As	442		99.90	NA	Mavrov et al. (2006)
	Se	2.32		98.70	NA	
Ultrafiltration	Cu(II)	78.74	8.5-9.5	100	NA	Juang et al. (2000)
	Zn(II)	81.10	8.5-9.5	95	NA	
	Cr(III	200	6.0	95	NA	Aliane et al. (2001)
Nanofiltrattion	Ni(II)	2000	3-7	94	NA	Ahn et al. (1999)
Reverse osmosis	Cu(II)	200	4-11	99	NA	Mohammad et al. (2004)
	Cd(II)	200	4-11	98	NA	
Froatation	Cu(II)	3.5	5.5	98.26	NA	Rubio et al. (1997)
	Ni(II)	2.0	5.5	98.6	NA	
	Zn(II)	2.0	5.5	98.6	NA	
	Zn(II)	50	7-9	100	NA	Matis et al. (2004)
Electrodialysis	Ni(II)	11.72	NA	69	NA	Tzanetakis et al. (2003)
	Co(II)	0.84	NA	90	NA	
Membrane electrolysis	Cr(VI)	130	8.5	99.6	7.9x10^3	Martınez et al. (2004)
	Ni(II)	2000	5.5	90	4.2x10^3	Orhan et al. (2002)
Electrochemical precitipation	Cr(VI)	570-2100	4.5	99	20	Kongsricharoern and Polprasert (1996)
	Cr(VI)	215-3860	1.5	99.99	14.7-20	Kongsricharoern and Polprasert (1995)
	Ni(II)	40000	NA	85	3.43x10^3	Subbaiah et al. (2002)
Ion exchange	Ni(II)	100		90	NA	
	Cr(III)	100	3-5	100	NA	Rengaraj et al. (2001)
	Ni(II)	100		90	NA	Sapari et al. (1996)
	Cu(II)	100		100	NA	
	Cr(VI)	9.77	NA	100	NA	Kabay et al. (2003)

NA=Not Available

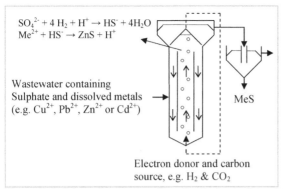

$$SO_4^{2-} + 4\,H_2 + H^+ \rightarrow HS^- + 4H_2O$$
$$Me^{2+} + HS^- \rightarrow ZnS + H^+$$

Wastewater containing
Sulphate and dissolved metals
(e.g. Cu^{2+}, Pb^{2+}, Zn^{2+} or Cd^{2+})

MeS

Electron donor and carbon
source, e.g. H_2 & CO_2

Figure 2-4: Schematic representation of a hydrogen-fed sulphate-reducing gas-lift bioreactor and settler treating wastewater polluted with heavy metals.

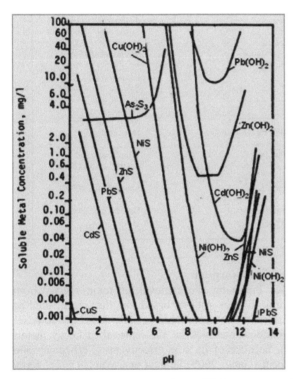

Figure 2-5: Comparison between the equilibrium concentrations of metal hydroxides and metal sulphides (Huisman et al., 2006)

Due to the low equilibrium concentrations of sulphide precipitates, most metals can be removed to very low concentrations (Figure 2-4) (Huisman et al., 2006). In addition, the high volumetric conversion rates (Table 2-6) allow for compact treatment systems. Moreover, the metal sulphide content of the produced sludge is high - 95% (Houten et al., 2006a). Often this metal sludge can be reused in the metallurgical industry (Huisman et al., 2006). The gas-lift bioreactor (Figure 2-3) is the most suitable bioreactor type for sulphate reduction with gaseous electron donors. In this system, the transfer of gas to the liquid is optimized and the sludge is partly retained in the bioreactor by a three phase separator. At zinc refinery in Budel (the Netherlands) and at the Kennecott copper mine (Utah, USA), hydrogen-fed sulphate-reducing gas-lift bioreactors are applied to remove metals from wastewater (Weijma et al., 2002; Houten et al., 2006a).

Table 2-6: Sulphate reduction rates achieved during different reactor runs with various electron donors.

e-donor	Temp (^{0}C)	Reactor concept	Volumetric activity (gSO_4^{2-} L^{-1} day^{-1})	Reference
Hydrogen	30	Gas-lift bioreactor	25	van Houten et al. (Van Houten et al., (2006b)
Hydrogen	30	Gas-lift bioreactor	30	van Houten et al. (1996)
Hydrogen	55	Gas-lift bioreactor	8	van Houten et al. (1997)
Hydrogen	30	Gas-lift bioreactor	13	van Houten et al. (1995a)
Hydrogen	30	Gas-lift bioreactor	5	Bijmans et al. (2008a)
Synthesis gas	30	Gas-lift bioreactor	7	van Houten et al. (1995b)
Synthesis gas	35	Packet bed reactor	1.2	du Preez et al. (1994)
CO	35	Packet bed reactor	2.4	du Preez et al. (1994)
CO	50-55	Gas-lift bioreactor	0.2	Sipma et al. (2007)
Formate	30	Membrane bioreactor	29	Bijmans et al. (2008b)
Methanol	65	Expanded granular sludge bed	15	Weijma et al. (2000)
Ethanol	35	Fluidized bed	5	Kaksonen et al. (2004)
Ethanol	8	Fluidized bed	0.6	Sahinkaya et al. (2007)
Ethanol	33	Expanded granular sludge bed	21	de Smul and Verstraete (1999)
Acetate	35	Packed bed	65	Stucki et al. (1993)
Acetate	33	Expanded granular sludge bed	10	Dries et al. (1998)
Molasses	30	Upflow anaerobic sludge bed	4.3	Annachhatre and Suktrakoolvait (2001)
Molasses	31	Packed bed	6.5	Maree and Strydom (1985)
Sucrose	55	Upflow anaerobic sludge bed	1	Lopes et al. (2007)
Sucrose	30	Upflow anaerobic sludge bed & Completely stirred tank reactor	1	Lopes et al. (2008)

2.2.4   Heavy metal sorption and biosorption

The search for new technologies involving the removal of toxic metals from wastewaters has directed attention to biosorption, based on the metal binding capacities of various biological materials. Biosorption has been demonstrated to poses a good potential to replace conventional methods for the removal of heavy metals (Vijayaraghavan and Yun, 2008), because of its cost effectiveness, efficiency and availability of biomass as biosorbent (Gadd, 2009; Fourest and Roux, 1992). Wang and Chen (2009) and Gadd and White (1993) defined biosorption as the removal of metal and metalloid species, compounds and particulates from solution by biological material. Ahluwalia and Goyal (2007) overviewed the major advantages and disadvantages of biosorption over conventional treatment of heavy metal contaminated wastewater. Natural and plant materials, industrial and agricultural wastes are receiving attention as a low cost sorption materials. Figure 2-6 shows a schematic diagram of a two-stage process for heavy metal by sorption.

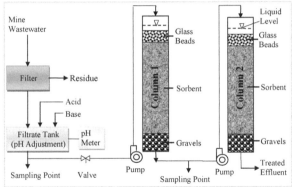

Figure 2-6: Schematic diagram of a two stage process for heavy metal removal by sorption

An overview of the sorption process for practical application is presented in Fig. 2-7 and 2-8. The hypothetical industrial plant (e.g. gold processing plant) operating round the clock generates 48, 000 US gallons per day (gpd) of wastewater containing 40 mg L^{-1} Cu, 30 mg L^{-1} Ni and 20 mg L^{-1} Zn at pH 6.5.

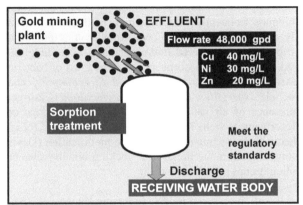

Figure 2-7: A hypothetical sorption gold mining effluent treatment plant

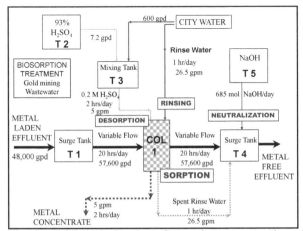

Figure 2-8: Schematic flowchart of a sorption process treating gold mining effluent treatment plant (modified from Volesky, 2003)

2.2.4.1   Sorption by microorganisms

Microorganisms include all prokaryotes (archaea and bacteria) and some eukaryotes (fungi and microalgae). Like all organic matter, microbial cells can take-up dissolved metals. However, the sorption capacities and affinities can differ greatly between non-microbial and microbial biomass and between the microorganisms themselves (Vijayaraghavan and Yun, 2008).

Two microbial metal-uptake processes are distinguished: bioaccumulation and biosorption. During bioaccumulation, the metals are accumulated intracellularly. For the transportation of metals across the cell membrane, the microorganisms need to be alive and metabolically active (Malik, 2004). Biosorption is the binding of metals to organic matter; it is a passive metabolic independent process. In the case of microorganisms, metals are mainly bound to functional groups on the outside of the cells wall. These functional groups include carboxyl, amine, phosphate and hydroxyl groups (van der Wal et al., 1997).

Ion exchange, electrostatic interaction and microprecipitation are examples of binding mechanisms that contribute to biosorption (Volesky and Schiewer, 1999; Han et al., 2004). Non viable microbial biomass frequently exhibits a higher affinity for metal ions than viable cells probably due to the absence of competing protons produced during metabolism. Sorption capacities of unaltered microorganisms exceeding 100 mg g^{-1} have often been observed. Potent metal biosorbents under the class of bacteria include species from the genera *Bacillus*, *Pseudomonas* and *Streptomyces*; and important fungal biosorbents include *Aspergillus*, *Rhizopus* and *Penicillium* (Castro et al., 2000). Due to the high presence of alginate in the cell wall, algae (singe or multicelluar) form excellent biosorbents as well. The sorption capacity of a microbial sorbent can be increased by chemical pretreatment or by genetic modification (Davis et al., 2003). Chemical pretreatment can remove impurities blocking binding sites or introduce new or enhanced binding groups into the cell wall.

In addition to sorption by the cell wall, metals can bind to inorganic compounds (e.g. sulphide, bicarbonate or phosphate) or extracellular polymeric substances (EPS), compounds excreted by metabolically active microorganisms.

2.2.4.2   Sorption using non-living biomass

Most of the biosorption studies have been and continue to be carried out on microbial systems; chiefly bacteria, microalgae and fungi, and with toxic metals and radionuclides (Gadd, 2009). However, practically all biological materials have an affinity for metal species. The process of biosorption using non-living biomass is a rapid phenomenon and involves a solid phase (sorbent or biosorbent; biological material) and a liquid phase (solvent, normally water) containing dissolved species to be sorbed (sorbate, metal ions). Due to the higher affinity of the sorbent for the sorbate species, the latter is attracted and bound there by different mechanisms (Beveridge et al., 1989). The degree of sorbent affinity for the sorbate determines its distribution between the solid and liquid phases.

2.2.4.3   Sorption using inorganic and natural sorbents

Both industrial by-products and natural materials locally available in certain regions can be employed as low-cost adsorbents. Due to its metal-binding capacity, natural materials such as zeolite and clay have been explored for treating metal-contaminated wastewater. Babel and Kurniawan (Babel and Kurniawan, 2003) studied Cr(VI) uptake from artificial wastewater using natural zeolite. According to them, NaCl treated zeolite had better removal capabilities (3.23 mg g^{-1}) for Cr(VI) ions than as-received zeolite (1.79 mg g^{-1}) at an initial Cr concentration of 20 mg L^{-1}. These results suggest that the Cr adsorption capacities of zeolite varied; depending on the extent of chemical treatment (Wingenfelder et al., 2005). The results were significantly lower than those of Peric´ et al. (2004) (Table 2-7). They also noted that metal removal by zeolite was a complex process, involving ion exchange and adsorption.

Another low-cost mineral that has a high cation exchange capacity (CEC) in solution is clay. There are three types of clay relevant for sorption: montmorillonite, bentonite and kaolinite. Out of the three, montmorillonite has the highest CEC (Virta, 2002). A number of studies on metal uptake using montmorillonite have been conducted. The adsorption of Cd(II), Cr(III), Cu(II), Ni(II) and Zn(II) ions on Na-montmorillonite using column operations was investigated by Abollino et al. (2003). The results of this and other works are shown in Table 2-7.

Table 2-7: Heavy metal uptake by natural materials through sorption

Sorbent	Metal	Initial Metal Conc. (mg L^{-1})	Adsorbent dosage (g L^{-1})	pH	Adsorption Capacity (mg g^{-1})	Reference
Natural zeolite	Zn(II)	65.4	10	6.5	13.06	Peric et al. 92004)
Kaolinite	Cd(II)	200	20	6.0	3.04	Ulmanu et al. (2002)
	Cu(II)	200	20	6.0	4.47	
Betonite	Cd(II)	200	20	6.0	9.27	Ulmanu et al. (2002)
	Cu(II)	200	20	6.0	18.16	
Pyrite fines	Cr(VI)	100	20	5.0-6.5	10.00	Zouboulis and Kydros (1993)
Ball clay		100				Chantawaong et al. (2003)
	Ni(II)	100	20	6.0	0.41	
	Cu(II)	100	20	5.0	1.60	
	Cd(II)	100	20	6.5	2.24	
	Zn(II)	100	20	6.5	2.88	
	Cr(III)		20	4.0	3.60	
Na-montmorillonite	Cd(II)	112	1274	5.5	5.20	Abollino et al. (2003)
	Cr(III)	52	1274	5.5	5.13	
	Cu(II)	63.5	1274	5.5	3.04	
	Ni(II)	58.7	1274	5.5	3.63	
	Zn(II)	65.4	1274	5.5	3.61	

2.2.4.4   *Biosorbent for heavy metal removal*

Since all biological materials have affinity for metals, and indeed other pollutants, the kinds of biomass potentially available for biosorption purposes are enormous. All kinds of microbial, plant and animal biomass, and derived products, have received investigation in a variety of forms and in relation to a variety of substances (Volesky et al., 2003; Gadd, 2009). Strong biosorbent behaviour of certain micro-organisms towards metallic ions is a function of the chemical composition of the microbial cells. This type of biosorbent consists of dead and metabolically inactive cells. A common rationale for such studies is to identify highly-efficient biosorbents that are cost-effective, i.e. cheap. This will, in theory, provide new opportunities for pollution control, element recovery and recycling (Gadd, 2009). A large quantity of materials has been investigated extensively as biosorbents for the removal of heavy metals or organics. Tables 2-8, 2-9 and 2-10 summarize the research works on metal uptake by agricultural waste and plant materials, bacterial and algal biomass, and industrial waste sorbents respectively.

Cost is an important parameter for comparing the sorbent materials. A biosorbent is considered low cost if it requires little processing, is abundant in nature, or is a by-product or waste material from another industry (Bailey et al., 1999; Gadd, 2009). Some of the reported low cost biosorbents includes sawdust, bark/tannin-rich materials, coconut husk and shells, peat moss, seaweed/algae, dead biomass etc (Bailey et al., 1999). Perhaps research should employ those biomass types that are efficient, cheap, easy to grow or harvest. Attention should be given to biomass modifications and/ alteration of the bioreactor configuration and physicochemical conditions to enhance biosorption. When biomass undergoes chemical treatment, there are always chance that its performance may suffer (Volesky, 2007). Figures 2-8 shows generalised scheme for biomaterial processing for biosorption application.

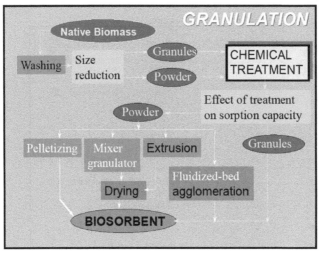

Figure 2-9: A generalised schematic flowchart of biomaterial processing into biosorbent granules with chemical treatment

From the above review, although abundant natural materials of cellulosic nature have been suggested as biosorbents, very little work has been actually done in this respect. Indeed, a large number of biological researches on heavy metal removal focused on the use of microorganisms such as bacteria. There is, therefore, the need to intensify research into the use of agricultural wastes and plant biomaterials such as coconut husk (CH), coconut shell (CS), and *Moringa oleifera seeds* (MOS) as low-cost and locally available biosorbents. Figure 2-9 and Figure 2-10 show pictures of coconut and *Moringa oleifera* biomaterials, respectively.

Table 2-8: Data on biosorption of heavy metals using different agricultural and plant biosorbents

Biosorbent	Metal	Initial Conc. (mg L^{-1})	Optimum pH	Temp (^{0}C)	Adsorption Capacity (mg g^{-1})	Reference
Calcinated wheat by-product	Cu(II)	10.0	6	20	8.40	Gherbi et al. (2004)
H$_3$PO$_4$-activated rubber wood sawdust	Cu(II)	20	6	30	3.75	Kalavathy et al. (2005)
Sawdust	Cu(II)	10	8	23	1.90	Yu et al. (2000)
Coconut coir	Cr(VI)				6.30	Gonzalez et al. (2008)
Submerged aquatic plant (*Myriophyllum spicatum*)	Cu(II) Pb(II) Zn(II)	10.00	5-6	25	10.37 46.49 15.59	Keskinkam et al. (2003)
Melon seed husk	Cd(II) Cu(II)	10-50	NA	29	23.30 11.40	Okieimen and Onyenkpa (1989)
Zr(IV)-loaded orange waste gel	As(III) As(V)		9-10 2-6	30	130.00 88.00	Biswas et al. (2008)
Agarose	Cu(II)	25-200	2	35	238.00	Anurag et al. (2007)
Coconut shell carbon (CSC)	Cr(VI)	20	6	NA	2.18	Babel and Kurniawan (2004)
HNO$_3$-Treated CSC	Cr(VI)	20	4	NA	10.88	Babel and Kurniawan (2004)
Sawdust	Cr(VI)	100	2	NA	15.82	Dakiky et al. (2002)
Banana peel	Cu(II) Zn(II)	25	6.8	NA	4.75 5.80	Annadurai et al. (2002)
Rice husk	Cd(II)	50	NA	NA	2.00	Ajmal et al. (2003)
Crab shell	Cu(II) Co(II)	500	6	NA	243.9 322.6	Vijayaraghavan et al. (2006)
Grape stalk	Cu(II) Ni(II)	5-300	5.2	20 ± 1	42.92 38.31	Vijayaraghavan et al. (2006)
Lignin	Pb(II) Cu(II) Cd(II) Zn(II) Ni(II)	0.2-2.5 0.2-2.5 0.2-2.5 0.2-2.5 0.2-2.5	5.5	20	1293.75 22.86 25.43 11.18 6.00	Guo et al. (2008)
Moringa	As(III) As(V)	1-100	2-12.5	NA	NA	Kumari et al. (2006)
Cassava waste	Cd(II) Cu(II) Zn(II)	112.4 63.5 65.4	4-5	NA	18.05 56.82 11.06	Abia et al. (2003)
Peanut hulls	Cu(II) Cd(II)	32 32	NA	NA	10.17 6.00	Brown et al. (2000)
Peanut hulls	Cu(II)	80	6-10	NA	65.57	Periasamy and Namasivayam (1996)
Cocoa shell	Cd(II) Cr(III) Cu(II) Ni(II) Zn(II)	28.10 13 15.9 14.7 16.4	2 2 2 2 2	NA	4.94 2.52 2.87 2.63 2.92	Meunier et al. (2003)
Bengal gram husk	Fe(III)	25	2.5	NA	72.16	Ahalya et al. (2006)
Pecan shell	Cu(II)	500	3.6		95.00	Shawabkeh et al. (2002)
Orange peel	Ni(II)	1000	6.0		158.00	Ajmal et al. (2000)
Fish scale	As(III) As(V)	20-100	4.0	20	2.48 2.67	Rahaman et al. (2008)
Barley straws	Cu(II) Pb(II)	10 10	NA	25 ± 1	4.64 23.20	Pehlivan et al. (2009)
Lessonia Nigrescens	As(V)	50-600	2.5		45.20	Hansen et al. (2005)
Sawdust	Cu(II)	10	6.3	23	7.50	Larous et al. (2005)
Unmodified coconut husk	As(III)	2000	5.0	30	944.00	Igwe and Abia (2007)
Modified coconut husk	As(III)	2000	5.0	30	952.5	Igwe and Abia (2007)

NA = Not Available

Table 2-9: Heavy metal uptake by bacterial and algal biomass

Bacterial/Algal species	Metal	Adsorption capacity (mg g^{-1})	Reference
Bacillus sp	Pb(II)	92.30	Tunali et al. (2006)
Streptomyces rimosus	Fe(III)	122.00	Selatnia et al. (2004)
Streptomyces rimosus	Zn(II)	80.00	Mameri et al. (1999)
Thiobacillus ferrooxidans	Zn(II)	82.60	Celaya et al. (2000)
Bacillus sp	Hg(II)	7.90	Green-Ruiz (2006)
Bacillus cereus	Pb(II)	36.71	
	Cu(II)	50.32	
Enterobacter sp	Cu(II)	32.50	Lu et al. (2006)
	Pb(II)	50.90	
	Cd(II)	46.20	
Pseudomonas putida	Cu(II)	96.90	Usla and Tanyol (2006)
	Pb(II)	270.40	
Pseudomonas stutzeri	Cu(II)	22.90	Nakajima et al. (2006)
Bacillus licheniformis	Cr(IV)	69.40	Zhou et al. (2007)
Bacillus thuringiensis	Cr(IV)	83.30	Sahin and Ozturk (2005)
Inonotus hispidus	As(III)	51.90	Sari and Tuzen (2009)
	As(V)	59.60	
Ulva lactuca	Pb(II)	34.70	Sari and Tuzen (2008)
	Cd(II)	29.20	
Spirogyra insignis	Cu(II)	19.30	Romera et al. (2006)
	Zn(II)	21.10	
	Pb(II)	51.50	
	Cd(II)	22.90	
	Ni(II)	17.50	
Spirogyra sp	Cu(II)	133.30	Gupta et al. (2006)
Ascophyllum nodosum	Cd(II)	215.00	Holan et al. (1993)

Table 2-10: Heavy metal uptake by industrial waste through sorption

Sorbent	Metal	Initial Metal Conc. (mg L^{-1})	Adsorbent dosage (g L^{-1})	pH	Adsorption capacity (mg g^{-1})	Reference
Sewage sludge	Cd(II)	40	10	5.5-6.0	16.00	Zhai et al. (2004)
Olive cake	Cr(VI)	100	2.4	2.0	33.44	Dakiky et al. (2002)
Olive cake	Cd(II)	100	NA	2.0-11.0	65.36	Al-Anber (2008)
Bagasse	Cd(II)	14	10	6.0	2.00	Gupta et al. (2003)
Fry ash	Ni(II)	12	10	6.5	1.70	Gupta et al. (2003)
Fish scale	As(III)	20-100	NA	4.0	2.48	Rahaman et al. (2008)
	As(V)	20-100			2.67	
Iron slags	Cu(II)	200	2	3.5-8.5	88.50	Feng et al. (2004)
	Pb(II)	200	2	3.5-8.5	95.24	
Red mud	Ni(II)	400	7	9.0	160.00	Zouboulis and Kydros (1993)
Soya cake	Cr(VI)	44	9.3	1.0	0.28	Daneshvar et al. (2002)
Anaerobic sludge	Cd(II)	200	20	6.0	5.82	Ulmanu et al. (2002)
Brewery waste	Pb(II)	33.1-1656	NA	4	85.49	Chen and Wang (2008)
	Ag(II)	17.3-864	NA	4	42.77	
	Sr(II)	14.1-704	NA	4	72.29	
	Cs(II)	21.3-1064	NA	4	10.11	
Tea factory waste	Cr(VI)	50-400	NA	2.0-5.0	54.65	Malkoc et al. (2007)

NA= Not Available

Figure 2-10: Figure 2-9 Coconut biomaterials

Figure 2-11: Moringa oleifera biomaterials

2.2.4.5   Sorption mechanisms

Biosorption is made possible by the ability of biological materials to accumulate heavy metals from wastewater through metabolically mediated or physicochemical uptake pathways. Due to the interaction of several factors on specific biosorbents, it is almost impossible to propose a general mechanism. Although, several metal-binding mechanisms have been put forward, e.g. physical and chemical adsorption, ion exchange and microprecipitation, the actual mechanism of metal biosorption is still not fully understood (Arief et al., 2008; Wang and Chan, 2008).

Chemisorption mechanism

Ion-exchange is an important concept in biosorption, because it explains many observations made during heavy metal uptake experiments (Davis et al., 2003). Ion-exchange is a reversible chemical reaction where an ion within a solution is replaced by a similarly charged ion attached onto an immobile solid particle (Han et al., 2006). Romero-Gonzalez et al. (2001) claimed that cadmium biosorption on *Saccharomyces cerevisiae* followed an ion-exchange mechanism. Tan and Cheng (2003) examined the mechanism involved in the removal of five heavy metals, namely, Cu(II), Ni(II), Zn(II), Pb(II) and Cr(III) by *Penicillium chrysogenum,* and concluded that ion-exchange was the dominant mechanism.

Ahmady-Asbchin et al. (2008) noticed that in the biosorption of copper ions on *Fucus serratus*, the bond between the copper ions and the surface functional groups of the biomass formed as soon as calcium ions were released from the surface. This ion-exchange process is one of the well-known surface reactions, which represents a great degree of complexity, primarily due to its multi-species nature. Kuyucak and Volesky (1989) reported that ion-exchange plays a role in the biosorption of cobalt by marine algae (*Ascophyllum nodosum*). In bacterial biosorption, the bacterial cell wall is the first component that comes into contact with the metal, where the solute can be deposited on the surface or within the cell wall structure (Vijayaraghavan and Yun, 2008). Since the mode of solute uptake by dead or inactive cells is extracellular, the chemical functional groups of the cell wall play vital a role in biosorption.

Involvement of functional groups in metal binding

Another frequently encountered metal binding mechanism is chelation, which can be defined as a firm binding of metal ions with an organic molecule (ligand) to form a ring structure (Arief et al., 2008). Various functional groups including carboxylic, hydroxyl, sulphate, phosphate, amides and amino groups can be considered possible for sorption. Among these groups, the amino group is the most effective for removing heavy metals, since it does not merely chelate cationic metal ions but also absorbs anionic species through the electrostatic interaction or hydrogen bonding. As they are negatively charged and abundantly available, carboxyl groups actively participate in the binding of metal cations (Deng and Ting, 2005). Table 2-11 shows major functional groups involve in metal binding.

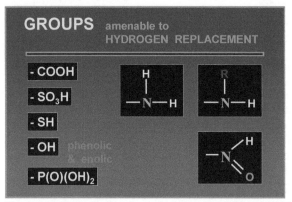

Figure 2-12: Major binding groups for biosorption (Volesky, 2007)

According to Ahalya et al. (2006), in the biosorption of Fe(III) by the husk of *Cider arientinum*, carboxyl as well as amino groups were involve in the metal uptake. Golag and Breitenbach (Golab and Breitenbach, 1995) indicated that carboxyl groups on the cell wall peptidoglycan of *Streptomyces pilosus* were responsible for the binding of Cu(II). The contribution of amine groups in the Cu(II) adsorption by *Mucor rouxii* was verified by Majumder et al. (2008). Altun and pehlivan (2007) revealed that chelation and ion-exchange were mainly behind Cu(II) adsorption from aqueous solution by walnut, hazelnut and almond shells.

Agricultural and plant waste materials, such as coconut husk and shell, are usually composed of lignin and cellulose as the main constituents. Other components present in the metal binding process are hemicellulose, lipids, proteins, simple sugars, starches, water, hydrocarbons and many more compounds that contain a variety of functional groups (Sud et al., 2008). The presence of these functional groups and their complexation with heavy metals during biosorption has been reported by different researchers using spectroscopic techniques (Tarley et al., 2004).

2.2.4.6   Sorption isotherms and models

Among all phenomena governing the mobility of substances in aqueous porous and aquatic environments, the transfer of substances from a mobile phase (liquid or gas) to a solid phase is a universal phenomenon. That is the reason why the 'isotherm', a curve describing the retention of a substance on a solid at various concentrations, is a major tool to describe and predict the mobility of this substance in the environment (Limousin et al., 2007). These retention/release phenomena are sometimes strongly kinetically controlled, so that time-dependence of the sorption isotherm must be specified.

Models have an important role in technology transfer from a laboratory scale to industrial scale (Limousin et al., 2007). The equilibrium of the sorption process is often described by fitting the experimental points with models (Table 2-10), usually used for the representation of isotherm sorption equilibrium. Appropriate models can help in understanding process mechanisms, analyze experimental data, predict answers to operational conditions and process optimization (Vijayaraghavan and Yun, 2008). As an effective quantitative means to compare binding strengths and design sorption processes, employing mathematical models for the prediction of the binding capacity can be useful (Volesky and Holan, 1995; Limousin et al., 2007). Examination and preliminary testing of the solid-liquid sorption system are based on two types of investigations: (a) equilibrium batch sorption tests and (b) dynamic continuous flow sorption studies. The two widely accepted and linearised equilibrium adsorption isotherm models for a single solute system are the Langmuir and the Freundlich isotherms (Table 2-12).

Surface complexation models provide molecular descriptions of metal adsorption using an equilibrium approach that defines surface species, chemical reactions, mass balances and charge balances. Such models provide information on stoichiometry and reactivity of adsorbed species (Goldberg and Crescenti, 2008). In order to use a surface complexation model, the adsorption mechanism and types of surface complexes must be specified for all adsorbing metal ions. This may necessitate independent experimental determination of the adsorption mechanism using spectroscopic techniques, including Raman and Fourier transform infrared (FTIR) spectroscopy, nuclear magnetic resonance (NMR) spectroscopy, electron spin resonance (ESR) spectroscopy, and X-ray absorption spectroscopy (XAS), which includes X-ray absorption near edge (XANES) and extended X-ray fine structure (EXAFS) spectroscopy, and X-ray reflectivity (Gadd, 2009). Most of these techniques have been used, but only in a small number of biosorption studies. Evaluation of equilibrium sorption performance needs to be supplemented with process-oriented studies of its kinetics and eventually by dynamic continuous flow tests.

Table 2-11: Frequently used single- and multi-component adsorption models

Models	Equation	Advantages	Disadvantages	References
Frequently used single-component adsorption models				
Langmuir	$$q = \dfrac{q_{max} b C_{eq}}{1 + b C_{eq}}$$	Interpretable parameters	Not structured: monolayer sorption	Pino et al. (2006)
Freundlich	$$q = K C_{eq}^{1/n}$$	Simple expression	Not structured: no leveling off	Amuda et al. (2007)
Combination (Langmuir – Freundlich)	$$q = \dfrac{q_{max} b C_{eq}^{1/n}}{1 + b C_{eq}^{1/n}}$$	Combination of the above equations	Unnecessarily complicated	Vijayaraghavan and Yun (2008)
Radke and Prausnitz	$$\dfrac{1}{q} = \dfrac{1}{a C_{eq}} + \dfrac{1}{b C_{eq}^{\beta}}$$	Simple expression	Empirical: uses three parameters	Gavrilescu (2004)
Reddlich-Petterson	$$q = \dfrac{a C_{eq}}{1 + b C_{eq}^{n}}$$	Approaches Freundlich at high concentrations	No special advantage	Gavrilescu (2004)
Brunnauer (BET)	$$q = \dfrac{BCQ^0}{(C_s - C)[1 + (B-1)C/C_s]}$$	Multilayer adsorption: inflection point	No "total capacity" equivalent	Gavrilescu (2004)
Dubinnin-Radushkevich	$$\dfrac{W}{W_0} = \exp\left[-k\left(\dfrac{\varepsilon}{\beta}\right)^2\right]$$	Temperature-independent	No limited behaviour in Henry's law regime	Gavrilescu 92004)
Frequently used multi-component adsorption model				
Langmuir (Multicomponent)	$$q_i = \dfrac{b_i q_{mi} C_i}{1 + \sum_{i=1}^{N} b_i C_i}$$			Goel et at. (2004)
Combination of Langmuir and Freundlich	$$q_i = \dfrac{a_i C_i^{1/n_i}}{1 + \sum_{i=1}^{N} b_i C_i^{1/n_i}}$$			Gavrilescu 92004)
IAST: Ideal Adsorbed Solution Theory	$$\dfrac{1}{q_t} = \sum \dfrac{Y_i}{q_t^0}$$			Limousin et al. (2007)
SCM: Surface Complexation Model	$$q \approx f(C_{eq})$$			Limousin et al. (2007)

NA: Not Available

2.2.4.7　　　*Parameters affecting sorption*

The investigation of factors affecting the efficiency of heavy metal sorption is of great interest for the industrial community. The efficiency is strongly influenced by the physico-chemical characteristics of the solutions, such as pH, temperature, initial metal concentration, presence of other ions and sorbent dosage (Arief et al., 2008). These factors are important in evaluating the maximal sorption performance of any sorbent.

pH

The solution pH is a crucial factor in heavy metal sorption. The pH value significantly influences the dissociation of the sorbent and the solution chemistry of the heavy metals, i.e. metal speciation, hydrolysis, complexation by organic and/or inorganic ligands, redox reactions and precipitation are all pH dependent (Fio et al., 2006; Akar et al., 2007; Arief et al., 2008; Guo et al., 2008; Ghodbane et al., 2008). Competition between cations and protons for binding sites means that sorption of metals like Cu, Cd, Ni, Co and Zn is often reduced at low pH values (Gadd and White, 1985; Gadd, 2009). Due to the importance of pH in sorption; many researches were conducted on its effect on the removal of various heavy metals.

Nuhoglu and Oguz (2003) reported that the removal of Cu(II) from aqueous solutions increases as the pH was increased, and that the maximum adsorption occurred at a pH of 7.7. Pino et al. (2006) noted that the removal of cadmium and arsenic by green coconut shell increased from 69% at pH 4 to 98% at pH 7. Cu(II) adsorption onto activated rubber and wood sawdust was found to be pH-dependent and maximum removal was observed at pH 6 (Virta, 2002). The optimum pH for As(III) removal (96.2%) by activated alumina was reported to be 7.6 (Singh and Pant, 2004). Yu et al. (2000) showed that the greatest increase in the sorption rate of Cu(II) on sawdust was observed in the pH range of 2 to 8. During the biosorption of Cu(II), Pb(II), Zn(II), and Cd(II) by Nile rose plant, Abdul-Ghani and Elchaghaby (2007) found out that the sorption efficiencies were pH-dependent, increasing by increasing the pH from 2.5 to 8.5, except for Pb(II). Other researchers, including Anurag et al. (2007), Pehlivan et al. (2009), Sari and Tuzen (2009) and Rahaman et al. (2008), also reported similar trends in the pH influence on biosorption of heavy metals such as copper, arsenic and lead ions.

Temperature

Depending on the structure and surface functional groups of a sorbent, temperature has an impact on the adsorption capacity within the range of 20-35 ^{0}C (Aksu et al., 1992; Veglio and Beolchini, 1997). It is well known that a temperature change alters the adsorption equilibrium in a specific way determined by the exothermic or endothermic nature of a process (Arief et al., 2008). Higher temperatures usually enhance sorption due to the increased surface activities and kinetic energy of the solute. However, physical damage can be expected at higher temperatures (Sag et al., 1998; Vijayaraghavan and Yun, 2007). It is always desirable to evaluate the sorption performance at room temperature, as this condition is easy to replicate.

The impact of temperature on the adsorption isotherm of Cu(II) and Cd(II) by corncob particles at a certain pH was explored by Shen and Duvnjak (Shen et al., 2004). They found out that the uptake of metal ions increased at a higher temperature. Igwe and Abia (2007) noted that temperature and particle size are very crucial parameters in biosorption reactions. They investigated the effects of these two parameters on the biosorption of As(III) from aqueous solution using modified and unmodified coconut fibers and found out that the most suitable temperature was 30 ^{0}C.

Ionic strength

Another important parameter in biosorption is the ionic strength, which influences the adsorption of solute to the biomass surface. Industrial wastewater often contains ions other than heavy metal ions, e.g., Na^+, K^+, Mg^{2+} and Ca^{2+}, which may interfere with heavy metal ion uptake by biomass. As a general trend, the metal uptake is found to decrease with increasing ionic strength of the aqueous solution as a result of more electrostatic attraction and change of the metal activity (Donmez and Aksu, 2002; Deng et al., 2007; Arief et al., 2008; Guo et al., 2008).

It was found that uranium uptake by biomass of bacteria, fungi and yeasts was not affected by the presence of manganese, cobalt, copper, cadmium, mercury and lead in solution (Sakaguchi and Nakajima, 1991). In contrast, the presence of Fe^{2+} and Zn^{2+} was found to influence uranium uptake by *Rhizopus arrhizus* (Dundar et al., 2008) and cobalt uptake by different microorganisms seemed to be completely inhibited by the presence of uranium, lead, mercury and copper.[188] Anions like CO_3^{2-} and PO_4^{2-} may clearly affect sorption through the formation of insoluble metal precipitates. Chloride may influence sorption through the formation of complexes, e.g. $CdCl_3^-$ (Gadd, 2009).

Metal concentration

The initial ion concentration can alter the metal removal efficiency through a combination of factors, i.e., availability of functional groups on the specific surface and the ability of surface functional groups to bind metal ions (especially at higher concentrations). Higher initial concentration of heavy metal ions results in a high solute uptake (Arief et al., 2008; Ho and Mckay, 2005). According to Arief et al. (2008), the initial concentration acts as a driving force to overcome mass transfer resistance to metal transport between the solution and the surface of the biomass.

A number of studies on the effect of the initial metal ion concentration have been undertaken in the past. Ajmal et al. (2005) found out that when the initial Cu(II) concentration is increased from 5 to 50 mg L^{-1}, the amount adsorbed increased from 8.8 to 96 mg L^{-1}, showing that adsorption of Cu(II) depends upon the initial concentration because the amount of Cu(II) adsorbed increased by increasing the initial concentration. Gulnaz et al. (2005) also investigated the effect of the initial Cu(II) ion concentrations between 100 and 400 mg L^{-1} at 20 ^{0}C and pH 4. They determined that the biosorption capacity of dried activated sludge for the < 0.063 mm particle size range was 76, 256 and 243 mg g^{-1} for 100, 200 and 400 mg L^{-1} initial Cu(II) concentrations, respectively. This indicated that the initial Cu(II) concentration is an important parameter for biosorption of Cu(II) by dried activated alumina. Anurag et al. (2007) also investigated the effect of initial concentration (25-200 mg L^{-1}) on biosorption of Cu(II) by agarose gels at 35 ^{0}C and reported that the removal was highest (70%) at an initial Cu(II) concentration of 25 mg L^{-1}. They explained that because at higher concentrations, more ions are competing for limited binding sites on the agarose gels, the rate of adsorption decreases, resulting in a lower adsorption percentage.

Adsorbent dosage

The amount of biomass in the solution also affects the specific metal uptake. For lower values of biomass concentrations, there is an increase in the specific uptake (Gadd et al., 1988; Fourest and Roux, 1992). Gadd et al. (1988) suggested that an increase in biomass concentration leads to interference between the binding sites. By increasing the adsorbent dosage, the adsorption efficiency increases even though the amount adsorbed per unit mass decreases. In principle, with more adsorbent present, the available adsorption sites or functional groups also increase. In turn, the amount of adsorbed heavy metal ions increase, which results in an improved adsorption efficiency (Mohanty et al., 2006; Amuda et al., 2007; Arief et al., 2008; Dundar et al., 2008; Gupta and Rastogi, 2008).

Adsorbent size

The size of the biosorbent also plays a vital role in biosorption. Smaller sized particles have a higher surface area, which in turns favours biosorption and results in a shorter equilibrium time. Simultaneously, a particle for biosorption should be sufficiently resilient to withstand the application of pressure and extreme conditions applied during regeneration cycles (Volesky, 2001). Therefore, preliminary experiments are necessary to determine the suitable size of a biosorbent.

2.2.4.8 Continuous-flow fixed bed sorption and desorption systems

An important observation from the literature reviewed so far is that most of the studies done were on laboratory scale batch sorption systems. Kratochvil and Volesky (1998) pointed out that the limited understanding of the metal uptake mechanism has hindered the application of biosorption. While batch equilibrium sorption studies can provide useful information on relative biosorbent efficiencies and important physico-chemical factors that affects biosorption, they usually provide no information on the removal mechanism. As a consequence, the batch method is very useful as a preliminary experiment, but extrapolation to porous or fixed bed media requires other investigations in a continuous system. Continuous mode of operation is preferred in large scale wastewater treatment applications due to advantages such as simple operation, high yield, easily scaled up from laboratory-scale procedure and easy regeneration of packed bed (Akar et al., 2009). Such laboratory continuous column studies which can provide information on the effect of flow rate variation and metal uptake mechanism by the biosorbents have received less attention.

Fixed bed sorption column process is the most effective configuration that allows continuous-flow (Volesky, 2003). It offers opportunity for virtually unlimited scale-up by using batteries of multiple columns that work in parallel and/or series to optimise the performance of the process. Fixed bed column makes optimum use of the concentration gradient between the solute sorbed by the solids and that remaining in the liquid phase. This provides the driving force for the sorption process. The process of metal biosorption in a fixed bed column in influenced by three key regimes: the sorption equilibrium (reaction), the sorption particle mass transfer and the flow pattern through the fixed bed (Crittenden and Weber, 1978; Liu and Weber, 1981). Combination of the three determines the overall performance of the sorption column which is judged by its service time, i.e. the length of time until the sorbed species "breaks through" the bed to be detected (at a given concentration) in the column effluent. At that point the bed is considered for all practical purposes "saturated" and the used up sorption column has to be taken out of operation for "regeneration". Table 2-13 shows data on biosorption of heavy metal in laboratory fixed-bed column (FBC) and continuous stirred tank reactor (CSTR).

In a continuous process, Pümpel et al. (2001) employed metal biosorbing or bioprecipitating bacteria to remove heavy metal from wastewater. The MEtal REmoval by Sand Filter INoculation (MERESAFIN) process is based on the inoculation of a sand filter with metal biosorbing or bioprecipitating bacteria. In this system, bacteria grow in a biofilm on a supporting material. During contact with heavy Zn^{2+}-containing wastewater, the biofilm in the extractive membrane bioreactor (EMB) - sulfate reducing bacteria (SRB) system adsorbs the metals. The metal-loaded biomass is then removed from the supporting material and the resting biomass residual on the substratum can be reused, after re-growth, for a subsequent treatment cycle. Chuichulcherm et al. (2001) studied the removal of Zn^{2+} from wastewater using SRB in an EMB. The continuous EMB-SRB system removed more than 90% (w/v) of the Zn^{2+} ions present in the wastewater. The concept is illustrated in Figure 2-11.

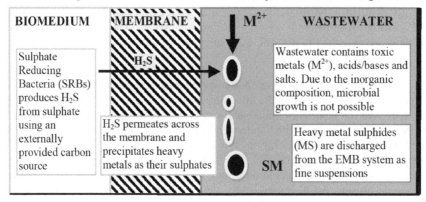

Figure 2-13: Concept of the EMB–SRB system (Modified from Chuichulcherm et al. (2001)).

It is also worth noting that considerable information is available on the biosorption of single-component systems, but many industries discharge effluents contain several components. Thus, knowledge of how one metal influences another is necessary for pilot reactor design. A successful laboratory scale continuous column bioreactor is a key to the design and operation of an on-site pilot scale reactor that uses the real gold mine wastewater; and ultimately, the design and operation of full scale industrial reactors based on low-cost agricultural/plant waste as biosorbent. Thus, current research must focus on column and pilot scale studies using multi-component metal ion solutions or actual industrial effluents.

One of the main attributes of biosorption is the potential ability to regenerate the biomass. However, most of the published works aimed to evaluate the binding ability of biomass and factors affecting the process. Less attention has been paid to the regeneration ability of the biosorbent, which often decides the industrial applicability of a process. Thus, biosorption studies should emphasise more on the ability of the biomass regeneration to improve the process viability. For operation of continuous flow systems, columns in parallel arrangements may allow sorption and desorption processes to occur without significant disruption.

Table 2-12: Data on heavy metal biosorption in fixed-bed column (FBC) and continuous stirred tank reactor (CSTR)

Biosorbent (Reactor type)	Metal	Initial ion conc (mg L^{-1})	pH	Temp (^0C)	Adsorbent dosage (g)	Packing height (cm)	Flow rate (ml min^{-1})	Uptake capacity (mg g^{-1})	R (%)	References
Silica-gel matrix (FBC)	Ni(II)	100	6.5	25±2	0.1	19	0.5-5.0	50.03	NA	Akar et al. (2009)
Seaweed (FBC)	Ni(II)	10	2.6	NA	NA	22.9	25.0	0.53	90	Brady et al. (1999)
	Zn(II)	10	2.6	NA	NA	22.9	25.0	0.35	90	
	Al(III)	10	2.6	NA	NA	22.9	25.0	0.97	74	
	Sb(II)	10	2.6	NA	NA	22.9	25.0	0.92	67	
Seaweed (*S. ilipendula*) (FBC)	Cu(II)	35	5	NA		41	15.0	38.0	NA	Volesky et al. (2003)
Poly acrylamide gel immobilized *S. platensis* (FBC)	Cu(II)	0.1	6	NA	2.0	NA	2.0	250	NA	Vannela and Verma (2006)
R. arrhizus (FBC)	Cr(VI)	100	2	25	NA	10	1.2	86.0	58	Sag et al. (2000)
	Fe(III)	100	2	25	NA	10	1.2	38.0	34	
Calcium-treated anaerobic biomass (FBC)	Pb(II)	40	4	NA	11.0	1.5	1.5	160	NA	Hawari and Mulligan (2006)
	Cu(II)	40	4	NA	11.0	1.5	1.5	91.0	NA	
	Cd(II)	40	4	NA	11.0	1.5	1.5	92.0	NA	
	Ni(II)	40	4	NA	11.0	1.5	1.5	53.0	NA	
Residual brewer *Saccharomyces cerevisiae* immobilized in volcanic rock (FBC)	Cr(III)	300	4.3		61.8	120	15.0	48.0	NA	Ramirez et al. (2007)
	Cr(VI)	200	1.7		61.8	120	15.0	60.0	NA	
Posidonia oceaica (FBC)	Cu(II)	1.02	6.0	22±1	0.9	10.0	0.7	41.5	NA	Izquierdo et al. (2009)
		1.98			1.0	9.8	0.7	45.0	NA	
		10.62			1.0	9.9	0.7	55.0	NA	
		20.07			1.0	10.1	0.7	56.0	NA	
		41.40			0.9	10.0	0.7	56.0	NA	
Sargassum fluitans seaweed biomass (FBC)	Cu(II)	2-3	2.5-3.5	NA		20	7.5	75	NA	Kratochvil et al. (1997)
Quaternized wood chips (FBC)	Cr(VI)	9.07	4.3	28±2	1.1	7.0	50.0	27.0	75	Low et al. (2001)
Cladosporium cladosporioides biomass beads (FBC)	Au(III)	10	4.0		3.0	10.0	0.13-0.4	110	80	Pethkar and Paknikar (1998)
Crab shell particles (FBC)	Ni(II)	100	4.5		31.0	15.0	5.0	25.0	60	Vijayaraghavan et al. (2004)
					41.0	20.0		25.0	68	
					51.0	25.0		26.0	72	
Olive stone (FBC)	Cr(III)	10	4.0	25	5.0	4.0	2.0	0.33	40	Calero et al. (2009)
		25			10.0	8.9	4.0	0.53	25	
		50			15.0	13.4	6.0	0.82	22	
Sargassum wightii	Cu(II)	100	4.5	NA	7.05	15	5	51.7	62	Vijayaraghavan and Prabu (2006)
		100			9.38	20	5	51.9	71	
		100			11.7	25	5	52.6	75	
		100			11.7	25	10	51.8	71	
		100			11.7	25	20	48.9	72	
		75			11.7	25	5	49.3	72	
		50			11.7	25	5	48.4	67	
Marine algae (*Gelidium*)	Cd(II)	25	4.7	20	10.7	15.0	4.1	19.0	NA	Vilar et al. (2008a)
Composite material (FBC)	Cd(II)	25	4.6	20	9.2	NA	4.0	7.90	NA	

NA= Not Available

Table 2-12: Continued

Biosorbent (Reactor type)	Metal	Initial ion conc (mg L^{-1})	pH	Temp (^{0}C)	Adsorbent dosage (g)	Packing height (cm)	Flow rate (mL min^{-1})	Uptake capacity (mg g^{-1})	R (%)	References
S. fruitans biomass (FBC)	Cu(II)	35.0	5.0	NA	33.0	38.0	10.0	61.5	NA	Kratochvil et al. (1995)
Hydrilla verticillata (FBC)	Cd(II)	10.0	5.0	25	0.5	NA	10.9	15.0	98	Bunluesin et al. (2003)
A. hydrophila (FBC)	Cr(VI)	104	1.5	30	13.3	19.0	2.0	44.6	79	Hasan et al. (2009)
Algae Gelidiun	Cd(II)	19.5	5.3	20	10	Not Applicable	35.5	12.4	NA	Vilar et al. (2008a)
Composite material (CSTR)	Cd(II)	20.4	5.4	20	10	Not Applicable	35.5	6.80	NA	
Algae Gelidiun	Cu(II)	25-6	4.3	NA	10	Not Applicable	35.5	53.9	NA	Vilar et al. (2008b)
	Pb(II)	25.6	4.8		10	Applicable	35.5	36.4		
Composite material (CSTR)	Cu(II)	25-6	4.3	NA	10	Not Applicable	35.5	16.0	NA	
	Pb(II)	25.6	4.8		10	Applicable	35.5	20.0		
Grape stalk (CSTR)	Cr(VI)	10	3	30	10	Not Applicable	NA			Escudero et al. (2008)

NA= Not Available

Desorption of loaded biomass enables reuse of the biomass, and recovery and/or containment of sorbed materials, although it is desirable that the desorbing agent does not significantly damage or degrade the biomass (Gadd and White, 1992; Gadd, 2009). In some cases, desorption treatments may further improve sorption capacities, although in other cases there may be a loss of sorption efficiency. A successful desorption process requires proper selection of eluents, which strongly depends on the type of biosorbent and the mechanism of biosorption. Also the eluent must be non-damaging to the biomass, less costly, environmentally friendly and effective.

A variety of substances have been used as metal desorbents including acids, alkalines and complexing agents depending on the substances sorbed, process requirements and economic considerations. Akar and Tunali (2005) studied the desorption of Cd(II) and Cu(II) using 10 mM HCl solution as desorbing agent. They reported that more than 95% of the adsorbed metal ions were desorbed from the biosorbent.

2.3 Conclusions

Pollution of the environment with heavy metals is widespread and often involves large volumes of wastewater. Remediation strategies for gold mining wastewater must be designed to support high throughput while keeping costs to a minimum. Biosorption is an alternative to traditional physicochemical means for removing toxic metals from wastewater.

Biosorption has a future as it can quickly (rapid intrinsic kinetics) and effectively sequester dissolved metals out of dilute complex solution with high efficiency. These characteristics make biosorption an ideal candidate for the treatment of high volumes of low concentration complex gold mine wastewater. However, preparing biosorption for application as a process requires a strong chemical engineering background, and an understanding of the sorption operation is a must.

It is therefore important to continue the fundamental research into the better understanding of the mechanism of biosorption. In this regard, it is necessary to focus current research on column studies to evaluate physicochemical conditions, which are necessary for pilot scale plant design and operations. Pilot scale studies are crucial for scaling up of the process to industrial level, which is the ultimate aim of all biosorption researches. The potential of plant materials and agricultural wastes as low-cost biosorbents should be exploited much more.

For industrial application of biosorption, regeneration of the biosorbents is important in keeping the process costs down and opening the possibility of recovering the metal ion extracted from the liquid phase.   Attention should be given to biomass modifications and alteration of bioreactor configurations to enhance biosorption. Research should employ those biomass types that are efficient, cheap, easy to grow or harvest. The successful design of hybrid reactors and regeneration of spent biosorbents is key to the commercialisation of biosorption.

2.4   Acknowledgements

The authors acknowledge funding from the Netherlands Fellowship Programme (NFP), Staff Development and Postgraduate Scholarship (Kumasi Polytechnic, Ghana) and the Unesco-IHE Partner Research Fund (UPaRF). We further acknowledge cooperation with AngloGold Ashanti (Obuasi, Ghana).

2.5   References

Abdel-Ghani, N.T., Elchaghaby, G.A., 2007. Influence of operating conditions on the removal of Cu, Zn, Cd and Pb ions from wastewater by adsorption. International Journal of Environmental Science and Technology 4 (4), 451-456.

Abia, A.A., Horsfall Jr, M., Didi, O., 2003. The use of chemically modified and unmodified cassava waste for the removal of Cd, Cu and Zn ions from aqueous solution. Bioresource Technology 90, 345-348.

Abollino, O., Aceto, M., Malandrino, M., Sarzanini, C., Mentasti, E., 2003. Adsorption of heavy metals on Na-montmorillonite. Effect of pH and organic substances. Water Research 37, 1619-1627.

Ahalya, A., Kanamadi, R.D., Ramachandra, T.V., 2006. Biosorption of iron(III) from aqueous solution using the husk of *Cicer arientinum*. Indian Journal of Chemical Technology 13, 122-127.

Ahluwalia, S.S., Goyal, D., 2007. Microbial and plant derived biomass for removal of heavy metals from wastewater. Bioresource Technology 98, 2243-2257.

Ahn, K.-H., Song, K.-G., Cha, H.-Y., Yeom, I.-T., 1999. Removal of ions in nickel electroplating rinse water using low-pressure nanofiltration. Desalination 122, 77-84.

Ajmal, M., Ali Khan Rao, R., Anwar, S., Ahmad, J., Ahmad, R., 2003. Adsorption studies on rice husk: removal and recovery of Cd(II) from wastewater. Bioresource Technology 86, 147-149.

Ajmal, M., Rao, R.A.K., Ahmad, R., Ahmad, J., 2000. Adsorption studies on *Citrus reticulata* (fruit peel of orange): removal and recovery of Ni(II) from electroplating wastewater. Journal of Hazardous Materials 79, 117-131.

Ajmal, M., Rao, R.A.K., Khan, M.A., 2005. Adsorption of copper from aqueous solution on *Brassica cumpestris* (mustard oil cake). Journal of Hazardous Materials 122, 177-183.

Akar, T., Kaynak, Z., Ulusoy, S., Yuvaci, D., Ozsari, G., Akar, S.T., 2009. Enhanced biosorption of nickel(II) ions by silica-gel-immobilized waste biomass: Biosorption characteristics in batch and dynamic flow mode. Journal of Hazardous Materials 163, 1134-1141.

Akar, T., Tunali, S., 2005. Biosorption performance of *Botrytis cinerea* fungal by-products for removal of Cd(II) and Cu(II) ions from aqueous solutions. Minerals Engineering 18, 1099-1109.

Akar, T., Tunali, S., Çabuk, A., 2007. Study on the characterization of lead (II) biosorption by fungus *Aspergillus parasiticus*. Applied Biochemistry and Biotechnology 136, 389-405.

Aksu, Z., Sag, Y., Kutsal, T., 1992. The biosorpnon of copper by *C. vulgaris* and *Z. ramigera*. Environmental Technology 13, 579-586.

Al-Anber, Z.A., Matouq, M.A.D., 2008. Batch adsorption of cadmium ions from aqueous solution by means of olive cake. Journal of Hazardous Materials 151, 194-201.

Aliane, A., Bounatiro, N., Cherif, A.T., Akretche, D.E., 2001. Removal of chromium from aqueous solution by complexation – ultrafiltration using a water-soluble macroligand. Water Research 35, 2320-2326.

Altun, T., Pehlivan, E., 2007. Removal of Copper(II) Ions from Aqueous Solutions by Walnut-, Hazelnut- and Almond-Shells. CLEAN – Soil, Air, Water 35, 601-606.

Amuda, O.S., Giwa, A.A., Bello, I.A., 2007. Removal of heavy metal from industrial wastewater using modified activated coconut shell carbon. Biochemical Engineering Journal 36, 174-181. Source. Water Environment Research 73, 118-126.

Annadurai, A., Juang, R.S. Lee, D.J., 2002. Adsorption of heavy metals from water using banana and orange peels. Water Sci Technol 47 (1), 185-90.

Anurag Pandey, D.B.A.S., Lalitagauri, R., 2007. Potential of Agarose for Biosorption of Cu(II) In Aqueous System. Am J Biochem Biotechnol **3** (2), 55-59.

Arief, V.O., Trilestari, K., Sunarso, J., Indraswati, N., Ismadji, S., 2008. Recent Progress on Biosorption of Heavy Metals from Liquids Using Low Cost Biosorbents: Characterization, Biosorption Parameters and Mechanism Studies. CLEAN – Soil, Air, Water 36, 937-962.

Ayoub, G.M., Semerjian, L., Acra, A., El Fadel, M., Koopman, B., 2001. Heavy metal removal by coagulation with seawater liquid bittern. Journal of Environmental Engineering 127, 196-207.

Babel, S., Kurniawan, T.A., 2003. A research study on Cr(VI) removal from contaminated wastewater using natural zeolite. J Ion Exchange 14, 289-292.

Babel, S., Kurniawan, T.A., 2004. Cr(VI) removal from synthetic wastewater using coconut shell charcoal and commercial activated carbon modified with oxidizing agents and/or chitosan. Chemosphere 54, 951-967.

Bailey, S.E., Olin, T.J., Bricka, R.M., Adrian, D.D., 1999. A review of potentially low-cost sorbents for heavy metals. Water Research 33, 2469-2479.

Beveridge, T.C., Doyle, R.J., 1989. Metal ions and bacteria. Wiley Intersience, New York.

Bijmans, M.F.M., Dopson, M., Lens, P.N.L., Buisman, C.J.N., 2008a. Effect of Sulfide Removal on Sulfate Reduction at pH 5 in a Hydrogen fed Gas-Lift Bioreactor. Journal of Microbiology and Biotechnology 18, 1809-1818.

Bijmans, M.F.M., Peeters, T.W.T., Lens, P.N.L., Buisman, C.J.N., 2008b. High rate sulfate reduction at pH 6 in a pH-auxostat submerged membrane bioreactor fed with formate. Water Research 42, 2439-2448.

Biswas, B.K., Inoue, J.-i., Inoue, K., Ghimire, K.N., Harada, H., Ohto, K., Kawakita, H., 2008. Adsorptive removal of As(V) and As(III) from water by a Zr(IV)-loaded orange waste gel. Journal of Hazardous Materials 154, 1066-1074.

Boddu, V.M., Abburi, K., Talbott, J.L., Smith, E.D., Haasch, R., 2008. Removal of arsenic (III) and arsenic (V) from aqueous medium using chitosan-coated biosorbent. Water Research 42, 633-642.

Bose, P., Aparna Bose, M., Kumar, S., 2002. Critical evaluation of treatment strategies involving adsorption and chelation for wastewater containing copper, zinc and cyanide. Advances in Environmental Research 7, 179-195.

Brady, J.M., Tobin, J.M., Roux, J.-C., 1999. Continuous fixed bed biosorption of Cu^{2+} ions: application of a simple two parameter mathematical model. Journal of Chemical Technology & Biotechnology 74, 71-77.

Brown, P., Atly Jefcoat, I., Parrish, D., Gill, S., Graham, E., 2000. Evaluation of the adsorptive capacity of peanut hull pellets for heavy metals in solution. Advances in Environmental Research 4, 19 29.

Bunluesin, S., Kruatrachue, M., Pokethitiyook, P., Upatham, S., Lanza, G.R., 2007. Batch and continuous packed column studies of cadmium biosorption by *Hydrilla verticillata* biomass. Journal of Bioscience and Bioengineering 103, 509-513.

Calero, M., Hernáinz, F., Blázquez, G., Tenorio, G., Martín-Lara, M.A., 2009. Study of Cr (III) biosorption in a fixed-bed column. Journal of Hazardous Materials 171, 886-893.

Can, C., Jianlong, W., 2008. Investigating the interaction mechanism between zinc and *Saccharomyces cerevisiae* using combined SEM-EDX and XAFS. Applied Microbiology and Biotechnology 79, 293-299.

Castro, H.F., Williams, N.H., Ogram, A., 2000. Phylogeny of sulfate-reducing bacterial. FEMS Microbiology Ecology 31, 1-9.

Celaya, R.J., Noriega, J.A., Yeomans, J.H., Ruiz-Manriquez, A., Ortega, L.J., 2000. Biosorption of Zn(II) by *Thiobacillus ferrooxidans*. Journal Name: Bioprocess Engineering 22 (6), 539-542.

Chantawong, V., Harvey, N.W., Bashkin, V.N., 2003. Comparison of heavy metal adsorptions by Thai kaolin and ball clay. Water Air Soil Pollut 148, 111–25.

Charerntanyarak, L., 1999. Heavy metals removal by chemical coagulation and precipitation. Water Science and Technology 39, 135-138.

Chuichulcherm, S., Nagpal, S., Peeva, L., Livingston, A., 2001. Treatment of metal-containing Chemical Technology & Biotechnology 76, 61-68.

Colleran, E., Finnegan, S., Lens, P., 1995. Anaerobic treatment of sulphate-containing waste streams. Antonie van Leeuwenhoek 67, 29-46.

Dakiky, M., Khamis, M., Manassra, A., Mer'eb, M., 2002. Selective adsorption of chromium(VI) in industrial wastewater using low-cost abundantly available adsorbents. Advances in Environmental Research 6, 533-540.

Daneshvar, N., Salari, D., Aber, S., 2002. Chromium adsorption and Cr(VI) reduction to trivalent chromium in aqueous solutions by soya cake. Journal of Hazardous Materials 94, 49-61.

Davis, T.A., Volesky, B., Mucci, A., 2003. A review of the biochemistry of heavy metal biosorption by brown algae. Water Research 37, 4311-4330.

de Smul, A., Goethals, L., Verstraete, W., 1999. Effect of COD to sulphate ratio and temperature in expanded-granular-sludge-blanket reactors for sulphate reduction. Process Biochemistry 34, 407-416.

Deng, L., Zhu, X., Wang, X., Su, Y., Su, H., 2007. Biosorption of copper(II) from aqueous solutions by green alga *Cladophora fascicularis*. Biodegradation 18, 393-402.

Deng, S., Ting, Y.P., 2005. Polyethylenimine-Modified Fungal Biomass as a High-Capacity Biosorbent for Cr(VI) Anions: Sorption Capacity and Uptake Mechanisms. Environmental Science & Technology 39, 8490-8496.

Dönmez, G., Aksu, Z., 2002. Removal of chromium(VI) from saline wastewaters by *Dunaliella species*. Process Biochemistry 38, 751-762.

Dries, J., De Smul, A., Goethals, L., Grootaerd, H., Verstraete, W., 1998. High rate biological treatment of sulfate-rich wastewater in an acetate-fed EGSB reactor. Biodegradation 9, 103-111.

du Preez, L.A., Maree, J.P., 1994. Pilot-scale biological sulphate and nitrate removal utilizing producer gas as energy source. Water Sci Technol 30, 275-285.

Dundar, M., Nuhoglu, C., Nuhoglu, Y., 2008. Biosorption of Cu(II) ions onto the litter of natural trembling poplar forest. Journal of Hazardous Materials 151, 86-95.

Escudero, C., Fiol, N., Poch, J., Villaescusa, I., 2009. Modeling of kinetics of Cr(VI) sorption onto grape stalk waste in a stirred batch reactor. Journal of Hazardous Materials 170, 286-291.

Esposito, A., Pagnanelli, F., Lodi, A., Solisio, C., Vegliò, F., 2001. Biosorption of heavy metals by *Sphaerotilus natans*: an equilibrium study at different pH and biomass concentrations. Hydrometallurgy 60, 129-141.

Feng, D., van Deventer, J.S.J., Aldrich, C., 2004. Removal of pollutants from acid mine wastewater using metallurgical by-product slags. Separation and Purification Technology 40, 61-67.

Fiol, N., Villaescusa, I., Martínez, M., Miralles, N., Poch, J., Serarols, J., 2006. Sorption of Pb(II), Ni(II), Cu(II) and Cd(II) from aqueous solution by olive stone waste. Separation and Purification Technology 50, 132-140.

Fourest, E., Roux, J.-C., 1992. Heavy metal biosorption by fungal mycelial by-products: mechanisms and influence of pH. Applied Microbiology and Biotechnology 37, 399-403.

Gad, G.M., White, C., de Rome, L., 1988. Heavy metals and radionuclide uptake by fungi and yest. In: Norris, P,R., Kelly, D.P., (Eds), Biohyhrometallurgy Science and Technology Letters. Kew, Surry, UK, pp. 421-435 (1988).

Gadd, G.M., 2009. Biosorption: critical review of scientific rationale, environmental importance and significance for pollution treatment. Journal of Chemical Technology & Biotechnology 84, 13-28.

Gadd, G.M., White, C., 1985. Copper uptake by Penicillium ochro-chloron: influence of pH on toxicity and demonstration of energy dependent copper influx using protoplasts. J Gen Microbiol 131, 1875-1879.

Gadd, G.M., White, C., 1992. Removal of thorium from simulated acid process streams by fungal biomass: Potential for thorium desorption and reuse of biomass and desorbent. Journal of Chemical Technology & Biotechnology 55, 39-44.

Gadd, G.M., White, C., 1993. Microbial treatment of metal pollution - a working biotechnology? Trends in Biotechnology 11, 353-359.

Garg, U.K., Kaur, M.P., Garg, V.K., Sud, D., 2007. Removal of hexavalent chromium from aqueous solution by agricultural waste biomass. Journal of Hazardous Materials 140, 60-68.

Gavrilescu, M., 2004. Removal of Heavy Metals from the Environment by Biosorption. Engineering in Life Sciences 4, 219-232.

Ghana EPA, 2007. Annual Report: Industrial effluent monitoring: Environmental Quality Department, Environmental Protection Agency, Accra, Ghana.

Gherbi, N., Méniai, A.H., Bencheikh-Lehocine, M., Mansri, A., Morcellet, M., Bellir, K., Bacquet, M., Martel, B., 2004. Study of the retention phenomena of copper II by calcinated wheat by-products. Desalination 166, 363-369.

Ghodbane, I., Nouri, L., Hamdaoui, O., Chiha, M., 2008. Kinetic and equilibrium study for the sorption of cadmium(II) ions from aqueous phase by eucalyptus bark. Journal of Hazardous Materials 152, 148-158.

Goel, J., Kadirvelu. K., Rajagopal, C., 2004. Competitive sorption of Cu(II), Pb(II) and Hg(II) ions from aqueous solution using coconut shell-based activated carbon, Defence Research and Development Organization, Delhi, India, pp 257-273.

Golab, Z., Breitenbach, M., Jezierski, A., 1995. Sites of copper binding in *Streptomyces pilosus*. Water, Air, & Soil Pollution 82, 713-721.

Goldberg, S., Criscenti, L.J., 2007. Modeling Adsorption of Metals and Metalloids by Soil Components. In: Biophysico-Chemical Processes of Heavy Metals and Metalloids in Soil Environments, eds: Violante A., Huang, P.M., Gadd. G.M., John Wiley & Sons, Inc., pp. 215-264.

Gonzalez, M.H., Araújo, G.C.L., Pelizaro, C.B., Menezes, E.A., Lemos, S.G., de Sousa, G.B., Nogueira, A.R.A., 2008. Coconut coir as biosorbent for Cr(VI) removal from laboratory wastewater. Journal of Hazardous Materials 159, 252-256.

Granato, M., Gonçalves, M.M.M., Boas, R.C.V., Sant'Anna Jr, G.L., 1996. Biological treatment of a synthetic gold milling effluent. Environmental Pollution 91, 343-350.

Green-Ruiz, C., 2006. Mercury(II) removal from aqueous solutions by nonviable *Bacillus sp.* from a tropical estuary. Bioresource Technology 97, 1907-1911.

Gulnaz, O., Saygideger, S., Kusvuran, E., 2005. Study of Cu(II) biosorption by dried activated sludge: effect of physico-chemical environment and kinetics study. Journal of Hazardous Materials 120, 193-200.

Gündoğan, R., Acemioğlu, B., Alma, M.H., 2004. Copper (II) adsorption from aqueous solution by herbaceous peat. Journal of Colloid and Interface Science 269, 303-309.

Guo, X., Zhang, S., Shan, X.-q., 2008. Adsorption of metal ions on lignin. Journal of Hazardous Materials 151, 134-142.

Gupta, V.K., Gupta, M., Sharma, S., 2001. Process development for the removal of lead and chromium from aqueous solutions using red mud—an aluminium industry waste. Water Research 35, 1125-1134.

Gupta, V.K., Rastogi, A., 2008. Biosorption of lead from aqueous solutions by green algae *Spirogyra species*: Kinetics and equilibrium studies. Journal of Hazardous Materials 152, 407-414.

Gupta, V.K., Rastogi, A., Saini, V.K., Jain, N., 2006. Biosorption of copper(II) from aqueous solutions by Spirogyra species. Journal of Colloid and Interface Science 296, 59-63.

Han, R., Zou, L., Zhao, X., Xu, Y., Xu, F., Li, Y., Wang, Y., 2009. Characterization and properties of iron oxide-coated zeolite as adsorbent for removal of copper(II) from solution in fixed bed column. Chemical Engineering Journal 149, 123-131.

Han, X., Wong, Y.S., Tam, N.F.Y., 2006. Surface complexation mechanism and modeling in Cr(III) biosorption by a microalgal isolate, *Chlorella miniata*. Journal of Colloid and Interface Science 303, 365-371.

Hansen, H.K., Rojo, A, Oyarzun, C, Ottosen, L.M., Ribeiro, A., Mateus, E., 2005. Biosorption of arsenic by *Lessonia nigrescens* in wastewater by copper smelting. In: Evaluation and Management of Drinking water sources contaminated with arsenic. Vol. CD-ROM.

Hasan, S., Srivastava, P., Ranjan, D., Talat, M., 2009. Biosorption of Cr(VI) from aqueous solution using *A. hydrophila* in up-flow column: optimization of process variables. Applied Microbiology and Biotechnology 83, 567-577.

Hawari, A.H., Mulligan, C.N., 2006. Heavy metals uptake mechanisms in a fixed-bed column by calcium-treated anaerobic biomass. Process Biochemistry 41, 187-198.

Ho, Y.S., McKay, G., 2000. The kinetics of sorption of divalent metal ions onto sphagnum moss peat. Water Research 34, 735-742.

Holan, Z.R., Volesky, B., Prasetyo, I., 1993. Biosorption of cadmium by biomass of marine algae. Biotechnology and Bioengineering 41, 819-825.

Huisman, J.L., Schouten, G., Schultz, C., 2006. Biologically produced sulphide for purification of process streams, effluent treatment and recovery of metals in the metal and mining industry. Hydrometallurgy 83, 106-113.

Igwe, J.C., Abia, A.A., 2007. Studies on the effects of temperature and particle size on bioremediation of As(III) from aqueous solution using modified and unmodified coconut fiber. Global Journal of Environ Res 1 (1), 22-26.

Ilan, S., Cabuk, A., Filik, C., Caliskan, F., 2004. Effect of pretreatment on biosorption of heavy metals by fungal biomass. Trakya Univ J Sci 5(1), 11-17.

Izquierdo, M., Gabaldón, C., Marzal, P., Álvarez-Hornos, F.J., 2010. Modeling of copper fixed-bed biosorption from wastewater by *Posidonia oceanica*. Bioresource Technology 101, 510-517.

Jennings, S.R., Neuman, D.R., Blicker, P.S., 2008. Acid Mine Drainage and Effects on Fish Health and Ecology: A Review. Reclamation Research Group Publication, Bozeman, MT.

Juang, R.-S., Shiau, R.-C., 2000. Metal removal from aqueous solutions using chitosan-enhanced membrane filtration. Journal of Membrane Science 165, 159-167.

Kabay, N., Arda, M., Saha, B., Streat, M., 2003. Removal of Cr(VI) by solvent impregnated resins (SIR) containing aliquat 336. Reactive and Functional Polymers 54, 103-115.

Kaksonen, A.H., Franzmann, P.D., Puhakka, J.A., 2004. Effects of hydraulic retention time and sulfide toxicity on ethanol and acetate oxidation in sulfate-reducing metal-precipitating fluidized-bed reactor. Biotechnology and Bioengineering 86, 332-343.

Kalavathy, M.H., Karthikeyan, T., Rajgopal, S., Miranda, L.R., 2005. Kinetic and isotherm studies of Cu(II) adsorption onto H_3PO_4-activated rubber wood sawdust. Journal of Colloid and Interface Science 292, 354-362.

Kapoor, A., Viraraghavan, T., 1995. Fungal biosorption — an alternative treatment option for heavy metal bearing wastewaters: a review. Bioresource Technology 53, 195-206.

Keskinkan, O., Goksu, M.Z.L., Yuceer, A., Basibuyuk, M., Forster, C.F., 2003. Heavy metal adsorption characteristics of a submerged aquatic plant (*Myriophyllum spicatum*). Process Biochemistry 39, 179-183.

Kongsricharoern, N., Polprasert, C., 1995. Electrochemical precipitation of chromium (Cr^{6+}) from an electroplating wastewater. Water Science and Technology 31, 109-117.

Kongsricharoern, N., Polprasert, C., 1996. Chromium removal by a bipolar electro-chemical precipitation process. Water Science and Technology 34, 109-116.

Kratochvil, D., Fourest, E., Volesky, B., 1995. Biosorption of copper by *Sargassum fluitans* biomass in fixed-bed column. Biotechnology Letters 17, 777-782.

Kratochvil, D., Volesky, B., 1998. Advances in the biosorption of heavy metals. Trends in Biotechnology 16, 291-300.

Kratochvil, D., Volesky, B., Demopoulos, G., 1997. Optimizing Cu removal/recovery in a biosorption column. Water Research 31, 2327-2339.

Kurniawan, T.A., Chan, G.Y.S., Lo, W.-H., Babel, S., 2006. Physico–chemical treatment techniques for wastewater laden with heavy metals. Chemical Engineering Journal 118, 83-98.

Kuyucak, N., Volesky, B., 1989. Accumulation of cobalt by marine alga. Biotechnology and Bioengineering 33, 809-814.

Larous, S., Meniai, A.H., Lehocine, M.B., 2005. Experimental study of the removal of copper from aqueous solutions by adsorption using sawdust. Desalination 185, 483-490.

Lazaridis, N.K., Matis, K.A., Webb, M., 2001. Flotation of metal-loaded clay anion exchangers. Part I: the case of chromates. Chemosphere 42, 373-378.

Lens, P., Vallerol, M., Esposito, G., Zandvoort, M., 2002. Perspectives of sulfate reducing bioreactors in environmental biotechnology. Reviews in Environmental Science and Biotechnology 1, 311-325.

Limousin, G., Gaudet, J.P., Charlet, L., Szenknect, S., Barthès, V., Krimissa, M., 2007. Sorption isotherms: A review on physical bases, modeling and measurement. Applied Geochemistry 22, 249-275.

Lopes, S.I.C., Dreissen, C., Capela, M.I., Lens, P.N.L., 2008. Comparison of CSTR and UASB reactor configuration for the treatment of sulfate rich wastewaters under acidifying conditions. Enzyme and Microbial Technology 43, 471-479.

Lopes, S.I.C., Sulistyawati, I., Capela, M.I., Lens, P.N.L., 2007. Low pH (6, 5, and 4) sulfate reduction during the acidification of sucrose under thermophilic (55°C) conditions. Process Biochemistry 42, 580-591.

Low, K.-S., Lee, C.-K., Lee, C.-Y., 2001. Quaternized wood as sorbent for hexavalent chromium. Applied Biochemistry and Biotechnology 90, 75-87.

Lu, W.-B., Shi, J.-J., Wang, C.-H., Chang, J.-S., 2006. Biosorption of lead, copper and cadmium by an indigenous isolate *Enterobacter sp.* J1 possessing high heavy-metal resistance. Journal of Hazardous Materials 134, 80-86.

Mack, C., Wilhelmi, B., Duncan, J.R., Burgess, J.E., 2007. Biosorption of precious metals. Biotechnology Advances 25, 264-271.

Majumdar, S.S., Das, S.K., Saha, T., Panda, G.C., Bandyopadhyoy, T., Guha, A.K., 2008. Adsorption behavior of copper ions on *Mucor rouxii* biomass through microscopic and FTIR analysis. C olloids and Surfaces B: Biointerfaces 63, 138-145.

Malik, A., 2004. Metal bioremediation through growing cells. Environment International 30, 261-278.

Malkoc, E., Nuhoglu, Y., 2007. Potential of tea factory waste for chromium(VI) removal from aqueous solutions: Thermodynamic and kinetic studies. Separation and Purification Technology 54, 291-298.

Mameri, N., Boudries, N., Addour, L., Belhocine, D., Lounici, H., Grib, H., Pauss, A., 1999. Batch zinc biosorption by a bacterial nonliving *Streptomyces rimosus* biomass. Water Research 33, 1347-1354.

Maree, J.P., Strydom, W.F., 1985. Biological sulphate removal in an upflow packed bed reactor. Water Research 19, 1101-1106.

Martnez, S.A., Rodriguez, M.G., Aguolar, R., Soto G., 2004. Removal of chromium hexavalent from rinsing chromating waters electrochemical reduction in a laboratory pilot plant. Water Science and Technology 49(1), 115-122.

Matis, K.A., Zouboulis, A.I., Gallios, G.P., Erwe, T., Blöcher, C., 2004. Application of flotation for the separation of metal-loaded zeolites. Chemosphere 55, 65-72.

Meena, A.K., Kadirvelu, K., Mishra, G.K., Rajagopal, C., Nagar, P.N., 2008. Adsorptive removal of heavy metals from aqueous solution by treated sawdust (*Acacia arabica*). Journal of Hazardous Materials 150, 604-611.

Meunier, N., Laroulandie, J., Blais, J.F., Tyagi, R.D., 2003. Cocoa shells for heavy metal removal from acidic solutions. Bioresource Technology 90, 255-263.

Mohan, D., Pittman Jr, C.U., Bricka, M., Smith, F., Yancey, B., Mohammad, J., Steele, P.H., Alexandre-Franco, M.F., Gómez-Serrano, V., Gong, H., 2007. Sorption of arsenic, cadmium, and lead by chars produced from fast pyrolysis of wood and bark during bio-oil production. Journal of Colloid and Interface Science 310, 57-73.

Mohanty, K., Jha, M., Meikap, B.C., Biswas, M.N., 2006. Biosorption of Cr(VI) from aqueous solutions by *Eichhornia crassipes*. Chemical Engineering Journal 117, 71-77.

Monser, L., Adhoum, N., 2002. Modified activated carbon for the removal of copper, zinc, chromium and cyanide from wastewater. Separation and Purification Technology 26, 137-146.

Muyzer, G., Stams, A.J.M., 2008. The ecology and biotechnology of sulphate-reducing bacteria. Nat Rev Microbiol 6, 441- 454.

Nakajima, A., Yasuda, M., Yokoyama, H., Ohya-Nishiguchi, H., Kamada, H., 2001. Copper biosorption by chemically treated *Micrococcus luteus* cells. World Journal of Microbiology and Biotechnology 17, 343-347.

Nuhoglu, Y., Oguz, E., 2003. Removal of copper(II) from aqueous solutions by biosorption on the cone biomass of *Thuja orientalis*. Process Biochemistry 38, 1627-1631.

Okieimen, F.E., Onyenkpa, V.U., 1989. Removal of heavy metal ions from aqueous solutions with melon (*Citrullus vulgaris*) seed husks. Biological Wastes 29, 11-16.

Orhan, G., Arslan, C., Bombach, H., Stelter, M., 2002. Nickel recovery from the rinse waters of plating baths. Hydrometallurgy 65, 1-8.

Pehlivan, E., Altun, T., Parlayıcı, S., 2009. Utilization of barley straws as biosorbents for Cu^{2+} and Pb^{2+} ions. Journal of Hazardous Materials 164, 982-986.

Periasamy, K., Namasivayam, C., 1996. Removal of copper(II) by adsorption onto peanut hull carbon from water and copper plating industry wastewater. Chemosphere 32, 769-789.

Perić, J., Trgo, M., Vukojević Medvidović, N., 2004. Removal of zinc, copper and lead by natural zeolite - a comparison of adsorption isotherms. Water Research 38, 1893-1899.

Pethkar, A.V., Paknikar, K.M., 1998. Recovery of gold from solutions using *Cladosporium cladosporioides* biomass beads. Journal of Biotechnology 63, 121-136.

Pino, G.H., de Mesquita, L.M.S., Torem, M.L., Pinto, G.A.S., 2006. Biosorption of Heavy Metals by Powder of Green Coconut Shell. Separation Science and Technology 41, 3141-3153.

Postgate J.R., 1984. The sulphate-reducing bacteria. Cambridge University Press, Cambridge.

Pümpel, T., Ebner, C., Pernfuß, B., Schinner, F., Diels, L., Keszthelyi, Z., Stankovic, A., Finlay, J.A., Macaskie, L.E., Tsezos, M., Wouters, H., 2001. Treatment of rinsing water from electroless nickel plating with a biologically active moving-bed sand filter. Hydrometallurgy 59, 383-393.

Rahaman, M.S., Basu, A., Islam, M.R., 2008. The removal of As(III) and As(V) from aqueous solutions by waste materials. Bioresource Technology 99, 2815-2823.

Ramirez C, M., Pereira da Silva, M., Ferreira L, S.G., Vasco E, O., 2007. Mathematical models applied to the Cr(III) and Cr(VI) breakthrough curves. Journal of Hazardous Materials 146, 86-90.

Randall, J.M., Hautala, E., Waiss, Jr. A.C., 1974. Removal and recycling of heavy metal ions from mining and industrial waste streams with agricultural by-products. In: Proceedings of the Fourth Mineral Waste Utilization Symposium. Chicago.

Rengaraj, S., Yeon, K.-H., Moon, S.-H., 2001. Removal of chromium from water and wastewater by ion exchange resins. Journal of Hazardous Materials 87, 273-287.

Rich, G., Cherry, K., 1987. Hazardous waste treatment technologies. *Pudvan Publishers*, New York.

Ripley, E.A., Redmann, R.E., Crowder, A.A., 1996. Environmental Effects of Mining. St. Lucie Press, Delray Beach, Florida, USA.

Romera, E., González, F., Ballester, A., Blázquez, M.L., Muñoz, J.A., 2006. Biosorption with Algae: A Statistical Review. Critical Reviews in Biotechnology 26, 223-235.

Romero-González, M.E., Williams, C.J., Gardiner, P.H.E., 2001. Study of the Mechanisms of Cadmium Biosorption by Dealginated Seaweed Waste. Environmental Science & Technology 35, 3025-3030.

Rubio, J., Tessele, F., 1997. Removal of heavy metal ions by adsorptive particulate flotation. Minerals Engineering 10, 671-679.

Sa, Y., Kutsal, T., 1999. An overview of the studies about heavy metal adsorption process by microorganisms on the lab. scale in Turkey, in: Amils, R., Ballester, A. (Eds.), Process Metallurgy. Elsevier, pp. 307-316.

Sağ, Y., Ataçoğlu, I., Kutsal, T., 2000. Equilibrium parameters for the single- and multicomponent biosorption of Cr(VI) and Fe(III) ions on R. arrhizus in a packed column. Hydrometallurgy 55, 165-179.

Şahin, Y., Öztürk, A., 2005. Biosorption of chromium(VI) ions from aqueous solution by the bacterium Bacillus thuringiensis. Process Biochemistry 40, 1895-1901.

Sahinkaya, E., Özkaya, B., Kaksonen, A.H., Puhakka, J.A., 2007. Sulfidogenic fluidized-bed treatment of metal-containing wastewater at 8 and temperatures is limited by acetate oxidation. Water Research 41, 2706-2714.

Sakaguchi, T., 1991. Nakajima A, Accumulation of heavy metals such as uranium and thorium by microorgnisms. In: Smith, R.W., Misra, M., (Eds), Mineral Bioprocessing. The Minerals, Metal and Materials Society.

Sapari, N., Idris, A., Hamid, N.H.A., 1996. Total removal of heavy metal from mixed plating rinse wastewater. Desalination 106, 419-422.

Sarı, A., Tuzen, M., 2009. Biosorption of As(III) and As(V) from aqueous solution by macrofungus (*Inonotus hispidus*) biomass: Equilibrium and kinetic studies. Journal of Hazardous Materials 164, 1372-1378.

Sari, A., Tuzen, M., Citak, D., Soylak, M., 2007. Adsorption characteristics of Cu(II) and Pb(II) onto expanded perlite from aqueous solution. Journal of Hazardous Materials 148, 387-394.

Schneegurt, M.A., Jain, J.C., Menicucci, J.A., Brown, S.A., Kemner, K.M., Garofalo, D.F., Quallick, M.R., Neal, C.R., Kulpa, C.F., 2001. Biomass Byproducts for the Remediation of Wastewaters Contaminated with Toxic Metals. Environmental Science & Technology 35, 3786-3791.

Sekhar, K.C., Subramanian, S., Modak, J.M., Natarajan, K.A., 1998. Removal of metal ions using an industrial biomass with reference to environmental control. International Journal of Mineral Processing 53, 107-120.

Selatnia, A., Bakhti, M.Z., Madani, A., Kertous, L., Mansouri, Y., 2004. Biosorption of Cd^{2+} from aqueous solution by a NaOH-treated bacterial dead *Streptomyces rimosus* biomass. Hydrometallurgy 75, 11-24.

Semerjian, L., Ayoub, G.M., 2003. High-pH–magnesium coagulation–flocculation in wastewater treatment. Advances in Environmental Research 7, 389-403.

Sengil, I.A., Özacar, M., 2008. Biosorption of Cu(II) from aqueous solutions by mimosa tannin gel. Journal of Hazardous Materials 157, 277-285.

Shammas N.K. (Eds.), Physico-chemical Treatment Processes, vol. 3, pp103–140, Humana Press, New Jersey.

Shammas, N.K., 2004. Coagulation and flocculation, in: Wang, L.K., Hung, Y.T., Shawabkeh, R.A., Rockstraw, D.A., Bhada, R.K., 2002. Copper and strontium adsorption by a novel carbon material manufactured from pecan shells. Carbon 40, 781-786.

Shawabkeh, R.A., Rockstraw, D.A., Bhada, R.K., 2002. Copper and strontium adsorption by a novel carbon material manufactured from pecan shells. Carbon 40, 781-786.

Shen, J., Duvnjak, Z., 2004. Effects of Temperature and pH on Adsorption Isotherms for Cupric and Cadmium Ions in Their Single and Binary Solutions Using Corncob Particles as Adsorbent. Separation Science and Technology 39, 3023-3041.

Shukla, S.R., Pai, R.S., 2005. Adsorption of Cu(II), Ni(II) and Zn(II) on modified jute fibres. Bioresource Technology 96, 1430-1438.

Singh, T.S., Pant, K.K., 2004. Equilibrium, kinetics and thermodynamic studies for adsorption of As(III) on activated alumina. Separation and Purification Technology 36, 139-147.

Sipma, J., Begoña Osuna, M., Lettinga, G., Stams, A.J.M., Lens, P.N.L., 2007. Effect of hydraulic retention time on sulfate reduction in a carbon monoxide fed thermophilic gas lift reactor. Water Research 41, 1995-2003.

Srivastava, N.K., Majumder, C.B., 2008. Novel biofiltration methods for the treatment of heavy metals from industrial wastewater. Journal of Hazardous Materials 151, 1-8.

Stucki, G., Hanselmann, K.W., Hürzeler, R.A., 1993. Biological sulfuric acid transformation: Reactor design and process optimization. Biotechnology and Bioengineering 41, 303-315.

Subbaiah, T., Mallick, S.C., Mishra, K.G., Sanjay, K., Das, R.P., 2002. Electrochemical precipitation of nickel hydroxide. Journal of Power Sources 112, 562-569.

Sud, D., Mahajan, G., Kaur, M.P., 2008. Agricultural waste material as potential adsorbent for sequestering heavy metal ions from aqueous solutions – A review. Bioresource Technology 99, 6017-6027.

Tan, T., Cheng, P., 2003. Biosorption of metal ions with *Penicillium chrysogenum* Applied Biochemistry and Biotechnology 104, 119-128.

Teixeira Tarley, C.R., Zezzi Arruda, M.A., 2004. Biosorption of heavy metals using rice milling by-products. Characterisation and application for removal of metals from aqueous effluents. Chemosphere 54, 987-995.

Tsezos, M., Volesky, B., 1982. The mechanism of uranium biosorption by *Rhizopus arrhizus*. Biotechnology and Bioengineering 24, 385-401.

Tunali, S., Çabuk, A., Akar, T., 2006. Removal of lead and copper ions from aqueous solutions by bacterial strain isolated from soil. Chemical Engineering Journal 115, 203-211.

Tünay, O., Kabdaşli, N.I., 1994. Hydroxide precipitation of complexed metals. Water Research 28, 2117-2124.

Tzanetakis, N., Taama, W.M., Scott, K., Jachuck, R.J.J., Slade, R.S., Varcoe, J., 2003. Comparative performance of ion exchange membranes for electrodialysis of nickel and cobalt. Separation and Purification Technology 30, 113-127.

Ulmanu, M., Marañón, E., Fernández, Y., Castrillón, L., Anger, I., Dumitriu, D., 2003. Removal of Copper and Cadmium Ions from Diluted Aqueous Solutions by Low Cost and Waste Material Adsorbents. Water, Air, & Soil Pollution 142, 357-373.

US.PA, 2005. Comprehensive Environment Response, Compensation and Liability Act (CERCLA), USA.

Uslu, G., Tanyol, M., 2006. Equilibrium and thermodynamic parameters of single and binary mixture biosorption of lead (II) and copper (II) ions onto *Pseudomonas putida*: Effect of temperature. Journal of Hazardous Materials 135, 87-93.

Van der Bruggen, B., Vandecasteele, C., 2002. Distillation vs. membrane filtration: overview of process evolutions in seawater desalination. Desalination 143, 207-218.

van der Wal, A., Norde, W., Zehnder, A.J.B., Lyklema, J., 1997. Determination of the total charge in the cell walls of Gram-positive bacteria. Colloids and Surfaces B: Biointerfaces 9, 81-100.

van Houten, B.H.G.W. Microbial aspects of synthesis gas fed bioreactors treating sulfate and metal rich wastewater. PhD thesis, Wageningen University, the Netherlands (2006a).

van Houten, B.H.G.W., Roest, K., Tzeneva, V.A., Dijkman, H., Smidt, H., Stams, A.J.M., 2006b. Occurrence of methanogenesis during start-up of a full-scale synthesis gas-fed reactor treating sulfate and metal-rich wastewater. Water Research 40, 553-560.

van Houten, R.T., Elferink, S.J.W.H.O., van Hamel, S.E., Pol, L.W.H., Lettinga, G., 1995. Sulphate reduction by aggregates of sulphate-reducing bacteria and homo-acetogenic bacteria in a lab-scale gas-lift reactor. Bioresource Technology 54, 73-79.

van Houten, R.T., Pol, L.W.H., Lettinga, G., 1994. Biological sulphate reduction using gas-lift reactors fed with hydrogen and carbon dioxide as energy and carbon source. Biotechnology and Bioengineering 44, 586-594.

van Houten, R.T., van der Spoel, H., van Aelst, A.C., Hulshoff Pol, L.W., Lettinga, G., 1996. Biological sulfate reduction using synthesis gas as energy and carbon source. Biotechnology and Bioengineering 50, 136-144.

van Houten, R.T., Yun, S.Y., Lettinga, G., 1997. Thermophilic sulphate and sulphite reduction in lab-scale gas-lift reactors using H_2 and CO_2 as energy and carbon source. Biotechnology and Bioengineering 55, 807-814.

Vannela, R., Verma, S., 2006. Cu^{2+} Removal and recovery by *Spi SORB*: batch stirred and up-flow packed bed columnar reactor systems. Bioprocess and Biosystems Engineering 29, 7-17.

Veglio, F., Beolchini, F., 1997. Removal of metals by biosorption: a review. Hydrometallurgy 44, 301-316.

Vigneswaran, S., Ngo, H.H., Chaudhary, D.S., Hung, Y.-T., 2005. Physicochemical Treatment Processes for Water Reuse Physicochemical Treatment Processes, in: Wang, L.K., Hung, Y.-T., Shammas, N.K. (Eds.). Humana Press, pp. 635-676.

Vijayaraghavan, K., Jegan, J., Palanivelu, K., Velan, M., 2004. Removal of nickel(II) ions from aqueous solution using crab shell particles in a packed bed up-flow column. Journal of Hazardous Materials 113, 223-230.

Vijayaraghavan, K., Jegan, J., Palanivelu, K., Velan, M., 2004. Removal of nickel(II) ions from aqueous solution using crab shell particles in a packed bed up-flow column. Journal of Hazardous Materials 113, 223-230.

Vijayaraghavan, K., Palanivelu, K., Velan, M., 2006. Biosorption of copper(II) and cobalt(II) from aqueous solutions by crab shell particles. Bioresource Technology 97, 1411-1419.

Vijayaraghavan, K., Prabu, D., 2006. Potential of Sargassum wightii biomass for copper(II) removal from aqueous solutions: Application of different mathematical models to batch and continuous biosorption data. Journal of Hazardous Materials 137, 558-564.

Vijayaraghavan, K., Yun, Y.-S., 2006. Chemical Modification and Immobilization of *Corynebacterium glutamicum* for Biosorption of Reactive Black 5 from Aqueous Solution. Industrial & Engineering Chemistry Research 46, 608-617.

Vijayaraghavan, K., Yun, Y.-S., 2008. Bacterial biosorbents and biosorption. Biotechnology Advances 26, 266-291.

Vilar, V.J.P., Botelho, C.M.S., Boaventura, R.A.R., 2008b. Lead and copper biosorption by marine red algae *Gelidium* and algal composite material in a CSTR ("Carberry" type). Chemical Engineering Journal 138, 249-257.

Vilar, V.J.P., Santos, S.C.R., Martins, R.J.E., Botelho, C.M.S., Boaventura, R.A.R., 2008a. Cadmium uptake by algal biomass in batch and continuous (CSTR and packed bed column) adsorbers. Biochemical Engineering Journal 42, 276-289.

Virta, R., 2002. USGS mineral information, US geological survey mineral commodity summary [http://www.minerals.usgs. gov/minerals/pubs/commodity/clays/190496.pdf] (Accessed: 21/08/09).

Volesky, B., 2001. Detoxification of metal-bearing effluents: biosorption for the next century. Hydrometallurgy 59, 203-216.

Volesky, B., 2007. Biosorption and me. Water Research 41, 4017-4029.

Volesky, B., Holan, Z.R., 1995. Biosorption of Heavy Metals. Biotechnology Progress 11, 235-250.

Volesky, B., Schiewer, S., 1999. Biosorption of metals. In: Flickinger M, Drew SW (eds). Encyclopedia of Bioprocess Technology. Wiley Intersience, New York, 433–53.

Volesky, B., Weber, J., Park, J.M., 2003. Continuous-flow metal biosorption in a regenerable Sargassum column. Water Research 37, 297-306.

Volesky, B., Weber, J., Park, J.M., 2003. Continuous-flow metal biosorption in a regenerable Sargassum column. Water Research 37, 297-306.

Wahab Mohammad, A., Othaman, R., Hilal, N., 2004. Potential use of nanofiltration membranes in treatment of industrial wastewater from Ni-P electroless plating. Desalination 168, 241-252.

Wang, J., Chen, C., 2006. Biosorption of heavy metals by Saccharomyces cerevisiae: A review. Biotechnology Advances 24, 427-451.

Weijma, J., Copini, C.F.M., Buisman, C.J.N., Schultz, C.E., 2002. Biological recovery of metals, sulfur and water in the mining and metallurgical industry, In: Water Recycling and Recovery in Industry / Lens, P.N.L., Hulshoff Pol, L.W., Wilderer, P., Asano, T. - London, UK : IWA Publishing, 2002. - (Integrated Environmental Technology Series). - ISBN 1 84339 005 1, pp. 605-622.

Weijma, J., Stams, A.J.M., Hulshoff Pol, L.W., Lettinga, G., 2000. Thermophilic sulfate reduction and methanogenesis with methanol in a high rate anaerobic reactor. Biotechnology and Bioengineering 67, 354-363.

White, D.M., Schnabel, W., 1998. Treatment of cyanide waste in a sequencing batch biofilm reactor. Water Research 32, 254-257.

Wingenfelder, U., Hansen, C., Furrer, G., Schulin, R., 2005. Removal of Heavy Metals from Mine Waters by Natural Zeolites. Environmental Science & Technology 39, 4606-4613.

Yu, B., Zhang, Y., Shukla, A., Shukla, S.S., Dorris, K.L., 2000. The removal of heavy metal from
       aqueous solutions by sawdust adsorption - removal of copper. Journal of Hazardous Materials
       80, 33-42.
Zhai, Y., Wei, X., Zeng, G., Zhang, D., Chu, K., 2004. Study of adsorbent derived from sewage sludge
       for the removal of Cd^{2+}, Ni^{2+} in aqueous solutions. Separation and Purification Technology 38,
       191-196.
Zhou, M., Liu, Y., Zeng, G., Li, X., Xu, W., Fan, T., 2007. Kinetic and Equilibrium Studies of Cr(VI)
       Biosorption by Dead *Bacillus licheniformis* Biomass. World Journal of Microbiology
       and Biotechnology 23, 43-48.
Zouboulis, A.I., Kydros, K.A., 1993. Use of red mud for toxic metals removal: The case of nickel.
       Journal of Chemical Technology & Biotechnology 58, 95-101.

Chapter 3

3 Assessment of the effluent quality from a gold mining industry in Ghana

This chapter has been presented and published as:

Acheampong, M.A., Meulepas, R.J.W., Lens, P.N.L., 2011. Characterisation of the Process Effluent of AngloGold-Ashanti Gold Mining Company in Ghana. In: *Proceedings of the 12th International Conference on Environmental Science and Technology*, Rhodes Island, Dodecanese, Greece (8 -10 September 2011).

Acheampong, M.A., Paksirajan, K., Lens, P.N.L., 2013. Assessment of the effluent quality from a gold mining industry in Ghana. Environmental Science and Pollution Research 20, 3799-3811.

Abstract

The physical and chemical qualities of the process effluent and the tailings dam wastewater of AngloGold-Ashanti Limited, a gold mining company in Ghana, were studied from June to September, 2010. The process effluent from the gold extraction plant contains high amounts of suspended solids and is therefore highly turbid. Arsenic, copper and cyanide were identified as the major pollutants in the process effluent with average concentrations of 10.0, 3.1 and 21.6 mg L^{-1}, respectively. Arsenic, copper, iron and free cyanide (CN^-) concentrations in the process effluent exceeded the Ghana EPA discharge limits; therefore, it is necessary to treat the process effluent before it can be discharged into the environment. Principal component analysis of the data indicated that the process effluent characteristics were influenced by the gold extraction process as well as the nature of the gold-bearing ore processed. No significant correlation was observed between the wastewater characteristics themselves, except for the dissolved oxygen and the biochemical oxygen demand. The process effluent is fed to the Sansu tailings dam, which removes 99.9% of the total suspended solids and 99.7% of the turbidity; but copper, arsenic and cyanide concentrations were still high. The effluent produced can be classified as inorganic, with a high value of electrical conductivity. It was noted that, though the Sansu tailings dam stores the polluted effluent from the gold extraction plant, there will still be serious environmental problems in the event of failure of the dam.

Keyword: cyanide; heavy metals; physico-chemical characteristics; gold mine wastewater; tailings dam; principal component analysis

3.1   Introduction

In many developing countries, industries dispose of their effluents without adequate characterisation, quantification and treatment due to lack of adequate legislation and law enforcement, as well as due to economic and technological constraints (Monsser et al., 2002; Acheampong et al., 2010; Sekabira et al., 2010). An important step in the selection of a treatment system is to study the physical and chemical characteristics of the process effluent stream in question (Igbinosa and Okoh, 2009). The study of a wastewater stream's characteristics will help to identify contaminants present in the wastewater and provides information on the treatment required to meet the regulatory limits for discharge.

The importance of the mining industry in modern life is evident, providing both social and economic gains, though commonly at the expense of the environment (Mitchell, 2009; Acheampong et al., 2010). Mining activities are important sources of contamination of the environment (Mitchell, 2009). Heavy metals such as lead, mercury, arsenic and cadmium are often implicated in human toxicity and poisoning (Schneegurt et al., 2001; Pena et al., 2004). Some heavy metals, such as zinc, copper, chromium, iron and manganese are essential for metabolic functioning in small amounts, but, in elevated quantities, these elements become toxic (Sari et al., 2007; Mohan et al., 2007).

The effect of metal pollution is long-lasting, as these pollutants are non-biodegradable (Gupta et al., 2001; Shukla and Pai, 2005). Cyanide binds readily with gold, silver and metals, and this characteristic makes it a viable and profitable chemical for gold extraction (Zagury et al., 2004). After the precious metal has been extracted from the ore, the cyanide- and heavy metal-laden wastewater is discharged as process effluent into the tailings pond. Although used abundantly, cyanide is highly toxic to human, plant and animal life (Zagury et al., 2004; Kyoseva et al., 2009). Cyanide is lethal to humans at a concentration of 2% (Kyoseva et al., 2009).

The objective of this study is to characterise and quantify the process and tailings dam effluents of AngloGold Ashanti limited in Obuasi (Ghana), thereby generating reliable data for planning and design of an effluent treatment system. To do this, effluent from the processing plant and the tailings pond were collected and analysed to determine the levels of physico-chemical parameters, and the heavy metals and cyanide concentrations. Principal component analysis of the data was done to establish the pollutant sources. The effluent quality was then assessed to determine its level of compliance with the Ghana EPA standard (Ghana EPA, 2010) for industrial effluent disposal.

3.2   Materials and methods

3.2.1   Study site
3.2.1.1   *Geographical location*
The Obuasi mine has a concession area of about 633 km^2 with a topography which varies from gently undulating to distinctly hilly and mountainous. Currently, the company is only engaged in underground mining. Two main ranges cut across the area. The major one, the Sansu-Moinsi range, runs centrally from the south-west to the north-east along the line from Obuasi to Sansu and consists of a series of low peaks typically about 500 m above sea level separated by steeply dissected valleys and slopes of 30-60% or more. The Sansu tailings dam (Fig. 3-1) is surrounded by cocoa and oil palm plantations. There are five rivers around the Sansu tailings dam, but only two have links with the dam. The Sansu facility uses the hybrid spigot and pad docking system of deposition in stable embankments built by telescoping rises of lateritic material. Several constructed seepage ponds and bore holes are used as monitoring points for the detection of leakages from the tailings dam into the ground water. The environmental laboratory is about 8 km from the tailings embankment.

3.2.1.2   *Types of gold-bearing ore*
Two types of gold-bearing ores are mined from the Obuasi deposits, namely, quartz vein and sulphide ores (arsenopyrite), which occur along a zone of intense shearing and faulting within Precambrian greenstones. The quartz vein type of ore consists mainly of quartz with free gold in association with lesser amounts of various metals (www.anglogold.com, accessed: 11/06/2012). The arsenopyrite (FeAsS) is associated with gold, iron, zinc, lead and copper (Vaugn and Craig 1978). This gold ore is generally non-refractory and is characterized by the inclusion of gold in the crystal lattice of the sulphide minerals. The percentage composition is 34.3, 46.0 and 19.7% for Fe, As and S, respectively.

3.2.1.3   Ore processing and wastewater generation

The ore processing plant of the Obuasi mine is located about 6 km south-west of the Sansu tailings dam. The plant processes over 470,000 tonnes of ore per month. The processing steps consist of crushing, milling, flotation and stirred tank bio-oxidation of the flotation concentrates. The biological oxidation process employs bacteria to effect enzymatic and chemical changes in the sulphide minerals concentrated from the flotation process. The bacteria oxidise the minerals and render them amenable to conventional cyanide leaching and activated carbon adsorption. At the end of the activated carbon adsorption process, the gold-loaded carbon is removed and washed before undergoing "elution" or desorption of gold cyanide at high temperature and pH. The bacteria gold recovery plant (called Biox) is the biggest in the world with a designed concentrate throughput of 960 tonnes per day (www.anglogold.com, accessed: 11/06/2012). The heavy metal ions and cyanide laden wastewater generated (i.e., the process effluent) is then pumped into the Sansu tailings pond for storage. The tailings dam covers about 2 km^2 of the Obuasi mine concession.

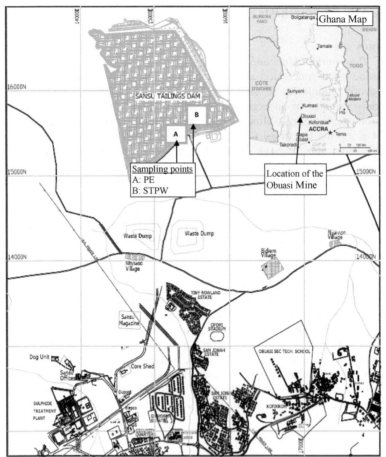

Figure 3-1: Map of the study area showing the Sansu tailings dam of the AngloGold Ashanti mine at Obuasi, Ghana. PE = Process effluent; STPW = Sansu tailings dam wastewater

3.2.1.4 *Wastewater treatment process of the Obuasi mine*

The processing plant discharges 286.2 tonnes h^{-1} of tailings and 400-500 m^3 h^{-1} of wastewater into the tailings dam. Some 500 m^3 h^{-1} of wastewater is pumped from the tailings pond for further treatment in a treatment plant (Figure 3-2), consisting of:

1. Two carbon-in-solution units, with design capacities of 300 m^3 h^{-1} each for the removal of cyanide, heavy metals and total suspended solids (TSS).

2. Ten rotating biological contactors having design capacities of 60 m^3 h^{-1} each for the treatment of cyanide only.

3. Chemical detoxification unit, which uses hypochlorite, lime and hydrogen peroxide to degrade cyanide at an influent flow rate of 100-200 m^3 h^{-1}.

4. Two Actiflo units, with design capacities of 320 m^3 h^{-1} each for the removal of cyanide, heavy metals (including arsenic), TSS and turbidity. The Actiflo unit is a compact, conventional-type water clarification system used in drinking water and wastewater applications that utilises microsand as a seed for floc formation. The microsand provides surface area that enhances flocculation and acts as a ballast or weight. The resulting sand-ballasted floc displays unique settling characteristics, which allow for clarifier designs with high overflow rates and short retention times. A portion of the treated effluent (about 80%) from the last pond is pumped to the processing plant for reuse during flotation, while the remaining is discharged into a nearby stream.

3.2.2 Wastewater sampling

Sampling, sample preparation, documentation and sampler cleaning were performed in accordance with procedures described in American Public Health Association (APHA) (1995). The plastic sample bottles were cleaned thoroughly using a detergent, 10 % HNO_3, triple-rinsed with distilled water and finally triple-rinsed with the sample. The wastewater samples were collected at a location of high turbulence to ensure good mixing. Samples were collected from the processing plant effluent discharge point and wastewater treatment plant intake point in the Sansu tailings dam (Figure 3-1).

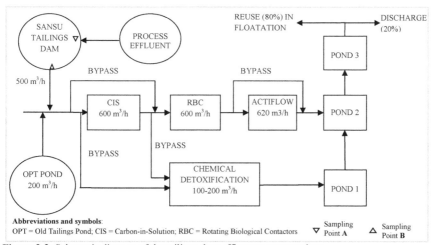

Figure 3-2: Schematic diagram of the tailings dam effluent treatment plant

The measuring campaign was between June 2010 and September 2010, during which samples were collected twice a day. The samples were transported to the environmental laboratory of the company in an ice chest containing ice cubes for analysis within 2 to 4 hours after collection. Samples for heavy metals analysis were filtered and acidified with 0.1 M HNO_3 before storing in a refrigerator. The sampling frequency was chosen such that the daily, weekly as well as monthly variations can be determined. The sampling period chosen was the transition period between the rainy and the dry seasons that provides average weather conditions during the study.

3.2.3   Wastewater characterisation
3.2.3.1   *Analytical quality assurance*
For all the methods that required the use of the spectrophotometer, both reagent blanks and sample blanks were used. The purpose of using blanks was to negate the effect of background interferences. When the absorbance was not zero for a particular parameter, the value was subtracted in order to obtain accurate determination of the parameter concerned. By blanking the instrument, it was assured that any reading obtained was exclusively due to the component of interest and not due to irrelevant chemicals in solution. Standard solutions were prepared for the chemical oxygen demand (COD) analysis. Before any measurement was done, instruments were calibrated using standard solutions. All field meters and equipment were checked and calibrated according to the manufacturers' specification. Preservation and handling of samples were done in accordance with procedures described in APHA (1995). All analyses were performed in triplicate.

3.2.3.2   Wastewater analysis
In order to characterise the wastewater, the following parameters were measured for the two sampling points: temperature, pH, electrical conductivity (EC), turbidity, TSS, total dissolved solids (TDS), dissolved oxygen (DO), sulphate (SO_4^{2-}), ammonium (NH_4^+), nitrate (NO_3^-), phosphate (PO_4^{3-}), COD, biochemical oxygen demand (BOD), heavy metal ions (i.e. arsenic (As), iron (Fe), copper (Cu), lead (Pb) and zinc (Zn)) and free cyanide (CN^-). Mercury (Hg), cadmium (Cd) and nickel (Ni) are not present in the effluent produced at the Obuai mine (AGA Obuasi 2010).

Temperature, pH, EC and TDS of the samples were determined onsite using a multi-parameter ion specific meter (Hanna instrument, combo). The DO was measured onsite with an oxygen meter (WTW field meter OXI 330). The turbidity was measured onsite using a Hanna turbidity meter, while the TSS was measured in the laboratory using a HACH spectrophotometer (DR 500) at a wavelength of 810 nm.

COD and BOD_5 were determined in the laboratory according to procedures described by the standard methods for the examination of water and wastewater (APHA, 1995). The concentrations of PO_4^{3-}, SO_4^{2-}, NO_3^- and NH_4^+ were determined in the laboratory using the HACH spectrophotometer (DR 5000) at wavelengths of 420, 450, 410 and 425 nm, respectively. The arsenic concentration in samples was determined by the graphite furnace method using an atomic absorption spectrometer (AAS). For heavy metal analysis of the samples, the direct air acetylene flame method using an AAS was used. The cyanide concentration was measured in the laboratory using Microquent cyanide test kits.

3.2.3.3   Statistical analysis

Standard statistical analysis was conducted using Excel and the Statistical Package for the Social Sciences (SPSS 17.0) software for windows (Boyacioglu, 2006; Liu et al., 2003). The SPSS software was used to perform descriptive statistical analysis (DSA) and principal component analysis (PCA), which is a very powerful technique applied to reduce the dimensionality of a data set consisting of a large number of interrelated variables, while retaining as much as possible the variability present in the data set (Zhang et al., 2009).

PCA reduces data complexity to a limited number of uncorrelated components that explain a large amount of variance (Jarboui et al., 2010). Such an approach facilitates the analysis by grouping data into smaller sets, preventing variable multi co-linearity problems. As a result, the variable relationships are determined allowing explanations of the data variance, while reducing the variable number to groups of individuals, based on principal component scores. Principal components were selected based on an eigenvalue greater than 1, a criterion for selecting the PC. The eigenvalue is the amount of variance explained by each factor, and factors with eigenvalues greater than 1 explain more total variation in the data than individual effluent characteristics (Das, 2009). PCA was interpreted in accordance with the hypothetical source of contaminants (i.e. geological, process, anthropogenic or a combination of all three). Varimax rotation was applied because orthogonal rotation minimises the number of variables with a high loading on each component and facilitates the interpretation of results (Mico et al., 2006).

3.3   Results

3.3.1   Variation in process and tailings dam effluents characteristics over time

The variations in the temperature recorded for the process effluent and the Sansu tailings dam effluent over the study period were minimal (Figure 3-3).

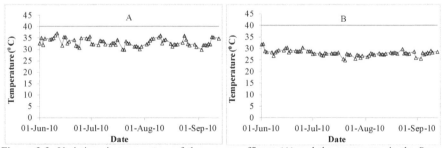

Figure 3-3: Variations in temperature of the process effluent (A) and the wastewater in the Sansu tailings dam (B) measured between June 2010 and September 2010, with Ghana EPA standard (-)

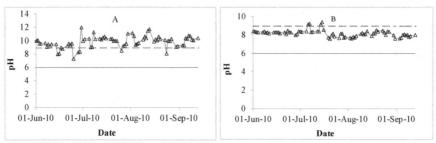

Figure 3-4: Variations in pH of the process effluent (A) and the wastewater in the Sansu tailings dam (B) measured between June 2010 and September 2010, with lower (-) and upper (---) limit in Ghana EPA standard

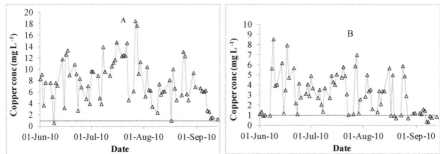

Figure 3-5: Variations in copper concentrations of the process effluent (A) and the wastewater in the Sansu tailings pond (B) measured between June 2010 and September 2010, with Ghana EPA Standard (-)

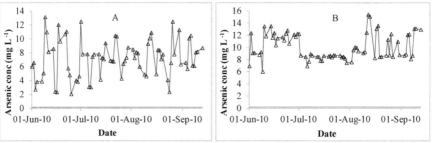

Figure 3-6: Variations in arsenic concentrations of the process effluent (A) and the wastewater in the Sansu tailings dam (B) measured between June 2010 and September 2010, with Ghana EPA Standard (-)

The pH values of the process effluent was found to be in most instances higher than the upper discharge limit of 9 (Figure 3-4A) (Ghana EPA, 2007). In contrast, the values measured for the Sansu tailings dam effluent were generally within the acceptable limit (Figure 3-4B). The heavy metal concentrations show significant variation during the study period (Figures 3-5 to 3-7). The concentration of cyanide in the process effluent was generally higher and showed more fluctuation than that of the tailings dam effluent (Figure 3-8).

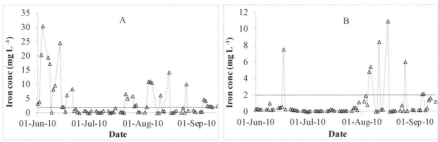

Figure 3-7: Variations in iron concentrations of the process effluent (A) and the wastewater in the Sansu tailings dam (B) measured between June 2010 and September 2010, with Ghana EPA Standard (-)

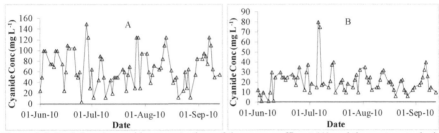

Figure 3-8: Variations in cyanide concentrations of the process effluent (A) and the wastewater in the Sansu tailings dam (B) measured between June 2010 and September 2010, with Ghana EPA Standard (-)

3.3.2   Physical characteristics of process and tailings dam effluents

The effluent from the gold processing plant contained very high solids concentrations, resulting in high values of EC, TDS, TSS, turbidity and COD (Table 3-1).

The 0.81% variation in the mean process effluent (PE) temperature between the morning and the afternoon (Table 3-1) indicated a proper control of the temperature during the gold extraction process. The pH values of the tailings dam wastewater were within the limit for safe discharge (WHO 1989; Ghana EPA 2010). The calculated average values for the morning and afternoon are 7.6 and 8.1, respectively. The Sansu tailings dam wastewater (STDW) showed a good buffering capacity as evident from the low standard deviations (Table 3-1).

The extremely high values for turbidity and TSS in the process effluent were due to the presence of high amounts of crushed rocks and earth materials from which the gold was extracted. The Ghana EPA standards for TSS and turbidity for wastewater are 50 mg L^{-1} and 75 NTU, respectively. In comparison, the tailings dam wastewater had much lower values for both TSS and turbidity (Table 3-1) due to TSS settling in the dam. The removal of TSS and turbidity from the effluent in the Sansu tailings dam is thus estimated to be 99.9 % and 99.7 %, respectively.

Table 3-1: Descriptive statistics of the process effluent characteristics for morning and afternoon taken daily for 75 days

Parameters	N	Minimum	Maximum	Mean (±SD)	Ghana EPA standards
Morning					
Temperature (^{0} C)	75	29.9	37.2	33.2 (±1.7)	40.0
DO (mg L^{-1})	75	0.7	7.7	4.6 (±1.1)	5.0
pH	75	7.5	12.0	9.8 (±0.7)	6.0 - 9.0
EC (µS cm^{-1})	75	3,390	5,700	4,560 (±433)	750
Turbidity (NTU)	75	658	45,700	16,400 (±9130)	75
TDS (mg L^{-1})	75	1,700	2850	2280 (±220)	50
TSS (mg L^{-1})	75	11,000	365,000	143,000 (±64000)	1,000
Cyanide (mg L^{-1})	75	4.0	150	66.3 (±32.5)	0.2
As (mg L^{-1})	75	2.1	13.2	7.2 (±2.7)	0.2
Fe (mg L^{-1})	75	0.0	30.5	3.8 (±6.1)	2.0
Cu (mg L^{-1})	75	0.6	18.5	7.8 (±3.9)	1.0
Pb (mg L^{-1})	75	0.0	1.4	0.2 (±0.2)	0.1
Zn (mg L^{-1})	75	0.0	2.4	0.1 (±0.3)	2.0
SO$_4^{2-}$ (mg L^{-1})	75	300	810	644 (±117)	250
NH$_4^+$ (mg L^{-1})	75	28.0	145	50.8 (±24.1)	1.5
NO$_3^-$ (mg L^{-1})	75	1.9	12.9	7.6 (±1.7)	11.5
PO$_4^{3-}$ (mg L^{-1})	75	2.0	30.8	13.5 (±5.8)	2.0
COD (mg L^{-1})	75	875	1,550	1,240 (±171)	250
BOD$_5$ (mg L^{-1})	75	0.4	7.4	4.3 (±1.1)	50.0
Afternoon					
Temperature (^{0} C)	75	30.4	37.2	33.5 (±1.5)	40.0
DO (mg L^{-1})	75	0.7	12.5	4.7 (±1.7)	5.0
pH	75	7.3	12.3	10.0 (±1.0)	6.0 - 9.0
EC (µS cm^{-1})	75	3,390	5,810	4,570 (±501)	750
Turbidity (NTU)	75	217	46,500	16,100 (±7,700)	75
TDS (mg L^{-1})	75	1,700	2,920	2300 (±257)	50
TSS (mg L^{-1})	75	8,600	267,000	140,000 (±59,400)	1,000
Cyanide (mg L^{-1})	75	4.0	150	66.7 (±33.5)	0.2
As (mg L^{-1})	75	1.7	13.2	7.1 (±2.6)	0.2
Fe (mg L^{-1})	75	0.0	26.2	4.5 (±6.6)	2.0
Cu (mg L^{-1})	75	1.0	18.4	8.0 (±3.9)	1.0
Pb (mg L^{-1})	75	0.0	1.6	0.2 (±0.3)	0.1
Zn (mg L^{-1})	75	0.0	2.8	0.2 (±0.5)	2.0
SO$_4^{2-}$ (mg L^{-1})	75	440	820	658 (±101)	250
NH$_4^+$ (mg L^{-1})	75	28.0	140	53.4 (±22.7)	1.5
NO$_3^-$ (mg L^{-1})	75	2.0	13.0	7.7 (±2.1)	11.5
PO$_4^{3-}$ (mg L^{-1})	75	2.5	29.4	13.3 (±5.6)	2.0
COD (mg L^{-1})	75	916	1,590	1,290 (±197)	250
BOD$_5$ (mg L^{-1})	75	0.4	11.9	4.3 (±1.6)	50.0

N = total number of samples analysed

3.3.3   Chemical characteristics of process and tailings dam effluents

The process effluent showed very high COD values throughout the study, with mean values of 1,240 and 1,290 mg L^{-1} for the morning and afternoon measurements (Table 3-1), respectively. These values are about five times higher than the permissible limit of the Ghana EPA (2007; 2010). The COD values for the tailings pond were lower than the Ghana EPA permissible limit of 250 mg L^{-1} in all cases analysed (Table 3-2). The BOD values were within the acceptable limit.

The DO concentration of the process effluent ranged from 0.7-7.7 and 0.7-12.0 mg L^{-1}, with mean values of 4.6 and 4.7 mg L^{-1} for the morning and afternoon samples (Table 3-1), respectively. The DO values for both the process and tailings pond (Table 3-2) effluents were lower than the minimum value of 5.0 mg L^{-1} set by the Ghana EPA (2007; 2010) and the World Health Organisation (WHO, 2008).

Table 3-2: Descriptive statistics of the tailings dam wastewater characteristics for morning and afternoon taken daily for 75 days

Parameters	N	Minimum	Maximum	Mean (±SD)	Ghana EPA standards
Morning					
Temperature (^{0}C)	75	25.0	32.1	28.2 (±1.2)	40.0
DO (mg L^{-1})	75	0.9	7.1	2.9 (±0.9)	5.0
pH	75	7.1	8.5	7.6 (±0.4)	6.0 - 9.0
EC (μS cm^{-1})	75	2,800	5,250	4,890 (±295)	750
Turbidity (NTU)	75	1.2	24.1	6.5 (±4.5)	75.0
TDS (mg L^{-1})	75	2,200	2,700	2,460 (±92.9)	50.0
TSS (mg L^{-1})	75	41	146	75 (±22)	1,000
Cyanide (mg L^{-1})	75	1.5	80.0	21.3 (±13.1)	0.2
As (mg L^{-1})	75	6.1	15.4	10.1 (±2.1)	0.2
Fe (mg L^{-1})	75	0.0	10.9	0.9 (±2.0)	2.0
Cu (mg L^{-1})	75	0.4	8.6	3.1 (±2.0)	1.0
Pb (mg L^{-1})	75	0.0	0.9	0.1 (±0.1)	0.1
Zn (mg L^{-1})	75	0.0	1.3	0.0 (±0.2)	2.0
SO_4^{2-} (mg L^{-1})	75	400	760	638 (±85.0)	250
NH_4^+ (mg L^{-1})	75	60.0	240	90.6 (±24.0)	1.5
NO_3^- (mg L^{-1})	75	0.9	6.8	3.2 (±1.0)	11.5
PO_4^{3-} (mg L^{-1})	75	2.1	10.6	6.7 (±2.0)	2.0
COD (mg L^{-1})	75	89	222	130 (±25)	250
BOD_5 (mg L^{-1})	75	0.9	6.6	2.6 (±0.8)	50.0
Afternoon					
Temperature (^{0}C)	75	25.7	32.2	28.8 (±1.5)	40.0
DO (mg L^{-1})	75	0.9	6.5	3.2 (±1.0)	5.0
pH	75	7.3	9.5	8.1 (±0.4)	6.0 - 9.0
EC (μS cm^{-1})	75	4,350	5,210	4,930 (±172)	750
Turbidity (NTU)	75	1.2	14.9	6.0 (±3.7)	75.0
TDS (mg L^{-1})	75	2,170	2,680	2,470 (±91.9)	50.0
TSS (mg L^{-1})	75	6	143	72 (±22)	1,000
Cyanide (mg L^{-1})	75	6.3	80.0	21.8 (±13.0)	0.2
As (mg L^{-1})	75	5.9	15.0	9.9 (±2.0)	0.2
Fe (mg L^{-1})	75	0.0	10.9	1.2 (±2.5)	2.0
Cu (mg L^{-1})	75	0.1	8.4	3.1 (±2.1)	1.0
Pb (mg L^{-1})	75	0.0	1.1	0.2 (±0.2)	0.1
Zn (mg L^{-1})	75	0.0	1.4	0.1 (±0.3)	2.0
SO_4^{2-} (mg L^{-1})	75	400	775	636 (±100)	250
NH_4^+ (mg L^{-1})	75	48.0	160	91.2 (±18.3)	1.5
NO_3^- (mg L^{-1})	75	0.9	5.5	3.1 (±0.9)	11.5
PO_4^{3-} (mg L^{-1})	75	2.0	18.2	6.5 (±2.2)	2.0
COD (mg L^{-1})	75	79	187	126 (±20)	250
BOD_5 (mg L^{-1})	75	1.0	6.3	2.9 (±1.0)	50.0

N = total number of samples analysed

For the inorganic compounds (NH_4^+, NO_3^-, PO_4^{3-} and SO_4^{2-}) analysed, only the NO_3^- average concentrations were below the Ghana EPA standards for industrial effluent (Table 3-1 and 3-2). Tables 3-1 and 3-2 show that the average NH_4^+ concentrations in the tailings pond effluent were lower than that of the process effluent in all cases analysed, while the average SO_4^{2-} remained comparatively the same in the process and tailings pond effluents.

Heavy metals (copper, arsenic, lead, iron and zinc) have been identified in the process and tailings pond effluent (Table 3-1). With the exception of arsenic, the mean copper, lead, iron and zinc concentrations in the tailings pond effluent were lower than those in process effluent (Tables 3-1 and 3-2). Tables 3-1 and 3-2 further showed that copper and arsenic concentrations exceeded the EPA permissible level in all the cases analysed.

The mean cyanide concentration in the process effluent measured during the morning and the afternoon were 66.3 and 66.7 mg L^{-1} (Table 3-1), respectively. Those recorded for the Sansu tailings dam effluent were 21.3 and 21.8 mg L^{-1} for the morning and afternoon (Table 3-2), respectively.

3.3.4   Principal components analysis
The data obtained during the study were subjected to PCA. The principal components of the 19 variables were generated and components extracted by PCA. Table 3-3 gives the calculated eigenvalue, per cent total variance, factor loading and cumulative variance for the process effluent and the tailings dam effluent. The PCA results were presented for the data sets obtained in the morning only, since the data sets obtained in the afternoon produced similar results (data not shown). PCA reduced the dimensionality from the 19 original parameters measured to six principal components (PCs) for each data set, explaining 66.4% (process effluent) and 68.3% (tailings dam effluent) of the data variability (Table 3-3).

3.4   Discussion

3.4.1   Process and tailings dam effluent characterisation
This study showed that most of the quality parameters of the process and the tailings dam effluent that were measured did not meet the permissible limits set by the Ghana EPA (Ghana EPA, 2007; 2010) for industrial effluent and the WHO guidelines for use of wastewater in agriculture and aquaculture (WHO, 1989). Apart from the unacceptably high levels of physico-chemical parameters such as TSS and TDS of the effluent produced, high arsenic, copper and cyanide concentrations of the effluent (Table 3-1 and 3-2) pose a serious environmental concern due to the toxicity of these elements. A summary of wastewater characteristics from different industries is presented in Table 3-4. It is difficult to make a direct comparison of the wastewater characteristics presented due to the dissimilar pH and temperature ranges of these different mineral-based industry effluents.

The temperatures were found to be below the Ghana EPA maximum permissible limit of 40 ^{0}C (Ghana EPA, 2007) throughout the study. The morning and the afternoon temperatures of the process effluent remained relatively stable during the study period, resulting in a fairly stable temperature of the Sansu tailings dam wastewater. This can be attributed to a well-controlled process temperature of the gold extraction plant. The lower mean tailings pond effluent temperature recorded during the morning and the afternoon compared with that of the process effluent can be attributed to the heat loss to the surrounding air and dissipation as a result of a larger surface area in the tailings dam.

The high pH recorded in the process effluent is attributed to the operational pH of approximately 12 used by the carbon-in-pulp/carbon-in-leach process in the gold extraction plant (www.anglogold.com, accessed: 11/01/2012). The Sansu tailings dam effluent pH gradually decreased over time due to neutralisation of the alkaline environment by rainwater and possibly also as a result of carbon dioxide uptake (Smith and Struhsacker, 1987; Zagury et al., 2004). The production of NO$_3^-$ and SO$_4^{2-}$ from the biodegradation of cyanide and thiocyanate (Gould et al., 2012) may have contributed to the lower pH value observed.

The Sansu tailings dam is an effective settler, as evident from the low levels of suspended solids and turbidity measured in the dam wastewater as compared with the levels found in the process effluent (Tables 3-1 and 3-2). The removal of TSS and turbidity in the tailings dam eliminates the need for filtration prior to treatment. The high values of the EC measured throughout the study period is due to the presence of high amounts of suspended solids and electrolytes, such as Fe^{3+}, Cu^{2+}, Pb^{2+}, SO_4^{2-}, PO_4^{3-} and NO_3^- in the process effluent (Table 3-1 and 3-2). The EC values of the tailings dam effluent were relatively higher than those of the process effluent in all cases, probably due to increased concentration as a result of evaporation loses in the tailings dam.

Wastewater containing biodegradable organic matter decreases the DO of the receiving water due to microbial oxygen consumption in the water. The average DO values obtained (Table 3-1 and 3-2) were below the minimum EPA standard for industrial effluent (Ghana EPA, 2010). The lower DO values recorded for the tailings dam wastewater (Table 3-1 and 3-2) compared with the process effluent (Table 3-1 and 3-2) during the study can be attributed to the poor aeration of the Sansu tailings dam, the low oxygen holding capacity of the wastewater in the dam as a result of the high tropical temperature (Sisodia and Moundiotiya, 2006) and the (bio)chemical oxidation of the sulphide ore minerals (Lundgren and Silver, 1980; Ahonen and Tuovinen, 1992).

The BOD and COD ratio indicates whether wastewater is toxic, biodegradable or stable (Samudro and Mangkoedihardjo, 2012). The ratio of the COD to BOD was more than 50, indicating low levels of biodegradable organic matter in the wastewater (Samudro and Mangkoedihardjo, 2010). Therefore, the high levels of COD recorded for the effluent may be attributed to the oxygen that is mainly required for the oxidation of both organic and inorganic compounds. The effluent produced can thus be classified as inorganic with a high load of non-biodegradable compounds.

Nitrate is the most oxidised form of nitrogen compounds in water and wastewater and causes algal blooming, resulting in eutrophication in surface water (Li et al., 2007; 2009). The tailings dam effluent shows low levels of nitrate (Table 3-2), but the ammonium, phosphate and sulphate concentrations, on the other hand, were higher than the Ghana EPA standards for both the process effluent and the tailings dam wastewater (Table 3-1 and 3-2).

Contamination of the effluent with heavy metals was the consequence of the processing method and the type of gold-bearing ore processed. Comparison of measured copper, arsenic, lead, iron and zinc concentrations to EPA standards for industrial effluent (Ghana EPA, 2010) indicates high copper and arsenic concentrations (Table 3-2). The higher arsenic concentration in the tailings dam effluent compared with the process effluent may partly be attributed to evaporation of water in the tailings dam. Furthermore, the arsenic concentration is also influenced by the pH of the effluent. At low pH, most of the arsenic will be adsorbed onto ferric hydroxide particles and settle to the bottom of the dam (Lucile et al., 2010). However, the pH of the Sansu tailings dam was not low enough to lead to the natural adsorption of arsenic onto the TSS and accumulation by TSS settling in the dam. Consequently, the soluble arsenic concentration in the liquid phase in the dam remained high, and far exceeded the EPA discharge limit for industrial effluent (Ghana EPA, 2010).

The cyanide contamination was due to its use as a process chemical. The lower concentrations of cyanide measured in the tailings dam effluent may be attributed to the natural removal processes, mainly volatilisation, complexation and biodegradation (Zagury et al., 2004). Naturally occurring microbial action causes transformation of cyanide to ammonia (NH_3). The cyanide reacts with the sulphide minerals and partially oxidise sulphur intermediates to produce thiocyanate (Mudder, 2001; Gould et al., 2012). Although the biodegradation of cyanide reduces the free cyanide concentration, it results in the formation of by-product, such as cyanate, thiocyanate, sulphate, ammonia, nitrate and elevated metal concentrations (Gould et al., 2012). Thiocyanate is less toxic and more stable than cyanide (Boucabeille et al., 1954). The rate of natural conversion of cyanide to NH_3 largely depends on environmental conditions and may not produce an effluent of desirable quality that meets discharge regulations (Lucile et al., 2010). In the case of the Sansu tailings dam at Obuasi, the continuous discharge of fresh tailings into the dam may impede the rate at which the natural cyanide transformations occur.

3.4.2   Pollutants source identification in the process effluent using PCA

The application of PCA to the data shows the loading of the wastewater characteristics on the principal components (Table 3-3), allowing to identify the sources of the contaminants in the wastewater. Loading values exceeding 0.71 are considered high, while values less than 0.32 are low (Liu et al., 2003; Yongming et al., 2006). Positive loading values indicate that the contribution of the variable increases with increasing loading in dimension, while negative loading values indicate a decrease (Maztlum et al., 1999).

For the data set pertaining to the process effluent, among the six significant PCs (Table 3-3), PC1 explained 15.2% of the total variance (Table 3-2) and has a strong positive loading on DO (0.92) and BOD_5 (0.93) and moderate negative loading on TSS (0.66), indicating the impact of geophysical source and organic matter on the effluent. Furthermore, the high loading for DO indicates the importance of oxygen to the biodegradation of organic matter and (bio)oxidation of reduced heavy metal sulphides present in the wastewater (An et al., 2010). Thus, PC1 is related to the source of the gold-bearing ore and bacteria involved in the (bio)oxidation of the ore.

PC2 explained 11.9% of the data variations and has a highly positive loading for EC (0.90) and TDS (0.91), reflecting the contribution of inorganic ions in the wastewater. This indicates the impact of geochemical origin of the ore and the gold extraction process on the wastewater quality. PC2 therefore showed a mixed source of contamination, that is, processing and geological. EC and TDS are therefore important physico-chemical parameters to be considered in the treatment of the effluent.

PC3 contributed 10.6% to the total variance explained, with high positive loading values for SO_4^{2-} (0.80) and cyanide (0.71), showing another contribution of both the processing and geochemical sources of contamination. PC3 identified the contaminants as originating from process-related and geological sources.

Table 3-3: Loading of measured wastewater characteristics (19) on principal components (rotated matrix [a]) for the process effluent (PE) and the Sansu tailings dam effluent (STDW)

Parameters	Principal components [b]					
	PC1	PC2	PC3	PC4	PC5	PC6
PE						
Temperature (^{0}C)	0.119	0.081	0.194	-0.268	0.035	0.140
DO (mg L^{-1})	**0.918**	0.094	-0.097	-0.011	-0.024	0.041
pH	0.171	-0.142	-0.076	**0.835**	0.027	0.152
EC (μS cm^{-1})	0.166	**0.897**	0.063	-0.033	0.082	0.056
Turbidity (NTU)	-0.390	0.254	0.333	-0.378	-0.035	0.171
TDS (mg L^{-1})	0.124	**0.905**	0.041	-0.035	0.047	0.126
TSS (mg L^{-1})	**-0.660**	-0.135	0.312	-0.009	0.008	0.174
Cyanide (mg L^{-1})	-0.174	0.245	**0.709**	0.092	-0.109	-0.056
As (mg L^{-1})	-0.145	0.076	0.098	**0.706**	-0.076	-0.394
Fe (mg L^{-1})	0.302	0.331	0.472	-0.073	**0.851**	-0.038
Cu (mg L^{-1})	-0.349	0.401	-0.325	0.022	**0.748**	-0.366
Pb (mg L^{-1})	0.081	-0.019	0.042	-0.051	**0.885**	0.000
Zn (mg L^{-1})	-0.111	0.150	0.010	0.028	**0.787**	-0.006
SO$_4^{2-}$ (mg L^{-1})	-0.169	-0.204	**0.797**	-0.021	0.080	0.122
NH$_4^+$ (mg L^{-1})	0.110	0.161	0.102	-0.076	-0.003	0.057
NO$_3^-$ (mg L^{-1})	0.045	0.074	0.141	-0.024	0.057	**0.847**
PO$_4^{3-}$ (mg L^{-1})	-0.351	0.184	0.294	0.547	-0.021	0.369
COD (mg L^{-1})	-0.247	0.169	-0.393	-0.004	-0.196	0.578
BOD$_5$ (mg L^{-1})	**0.925**	0.148	0.066	0.001	0.000	0.059
Eigenvalue	2.88	2.26	2.02	1.74	1.73	1.63
Total variance (%)	15.2	11.9	10.6	9.1	9.1	8.6
Cumulative variance (%)	15.2	27.1	37.7	46.8	55.9	64.5
STDW						
Temperature (^{0}C)	0.368	0.101	0.290	0.380	**0.605**	-0.262
DO (mg L^{-1})	0.191	-0.012	**0.954**	0.090	0.036	0.082
pH	**0.686**	-0.112	0.197	-0.272	0.040	-0.039
EC (μS cm^{-1})	-0.004	0.015	0.082	**0.793**	-0.110	0.060
Turbidity (NTU)	**0.779**	0.001	0.097	0.352	-0.116	-0.107
TDS (mg L^{-1})	0.042	-0.076	0.022	**0.848**	0.117	-0.086
TSS (mg L^{-1})	**0.624**	-0.088	0.184	0.349	-0.051	-0.283
Cyanide (mg L^{-1})	0.301	-0.034	0.086	-0.290	-0.066	0.494
As (mg L^{-1})	-0.148	-0.119	-0.190	0.124	**0.804**	0.029
Fe (mg L^{-1})	-0.166	**0.721**	0.042	0.046	0.071	0.227
Cu (mg L^{-1})	0.363	0.328	-0.268	-0.256	0.120	**0.881**
Pb (mg L^{-1})	0.152	**0.880**	-0.039	-0.135	0.112	-0.162
Zn (mg L^{-1})	-0.080	**0.909**	-0.037	0.021	-0.121	-0.051
SO$_4^{2-}$ (mg L^{-1})	-0.386	-0.067	0.249	0.155	-0.094	**0.735**
NH$_4^+$ (mg L^{-1})	**-0.604**	-0.034	0.087	0.016	-0.218	-0.143
NO$_3^-$ (mg L^{-1})	**0.736**	-0.033	0.153	-0.095	-0.192	-0.011
PO$_4^{3-}$ (mg L^{-1})	-0.009	-0.144	-0.170	0.257	**-0.800**	0.012
COD (mg L^{-1})	0.072	0.047	0.001	-0.037	0.024	0.035
BOD$_5$ (mg L^{-1})	0.082	-0.022	**0.957**	0.039	0.021	0.040
Eigenvalue	3.01	2.31	2.24	2.11	1.84	1.31
Total variance (%)	15.8	12.2	11.8	11.1	9.7	6.9
Cumulative variance (%)	15.8	28.0	39.8	50.9	60.6	67.5

[a] Rotation method: Varimax with Kaiser normalization
[b] Extraction method: principal components analysis

The contribution of PC4 to the variance is 9.1%. It has a high positive loading for pH (0.84) and As (0.71), showing the important relationship between pH and speciation of arsenic in solution. PC4 shows the impact of the gold-bearing ore (arsenopyrite) on the effluent quality. This suggests a geological source of contamination rather than the mineral ore processing.

The highly positive loading value of Fe (0.85), Cu (0.75), Pb (0.89) and Zn (0.79) on PC5 suggests a geological source of pollution as these heavy metals are associated with the arsenopyrite ore. These degrees of covariance strongly suggest a common source for the metals (Bai et al., 2011). The high loading value of Cu also suggests a contribution from the extraction process because $CuSO_4$ is used as a catalyst during the extraction process (www.anglogold.com, accessed: 11/06/2012). However, this contribution is negligible compared with the large amount of ore processed. PC6 on the other hand depicts the organic source of contamination associated with the ore and the earth materials processed.

3.4.3   Pollutant source identification in the tailings dam effluent using PCA

For the data set pertaining to the tailings dam wastewater, among the six PCs (Table 3-3), PC1 contributed 15.8% to the total variation observed and has high positive loading values for turbidity (0.78) and NO_3^- (0.74), moderate positive loading values for pH (0.69) and TSS (0.62) and moderate negative loading value for NH_3 (0.60), indicating mixed sources of contamination. This suggests that the contribution of turbidity, TSS, NO_3^- and pH will increase with increasing loading, while NH_4^+ decreases with increasing loading. PC1 confirms the direct relationship between turbidity and TSS.

PC2 has Fe (0.72), Pb (0.88) and Zn (0.91) as significant parameters with high positive loading values, indicating the contribution from the ore. The positive eigenvectors indicate an increasing contribution to the wastewater quality with pollutants loading. The high loading values for DO (0.95) and BOD_5 (0.96) on PC3 show the importance of oxygen to the biodegradation of organic matter in the wastewater. PC4 has high positive loading values for EC (0.79) and TDS (0.85), which is an indication of mixed sources of pollution. In particular, electrolytes in the wastewater (such as Cu^{2+}, Fe^{3+}, SO_4^{2-} and CN^-) contributed to this component.

Table 3-4: Summary of wastewater characteristics from different mineral processing based industrial sources

Wastewater source	Temp (°C)	pH	EC (μS cm⁻¹)	TSS (mg L⁻¹)	TDS (mg L⁻¹)	Copper (mg L⁻¹)	Arsenic (mg L⁻¹)	Iron (mg L⁻¹)	Reference
Copper mine	23.0	2.1	NA	NA	NA	18.6	NA	55.7	Islam-nd-din et al. (2010)
Copper mine	NA	4.8	NA	NA	NA	10.3	NA	120.0	Mahiroglu et al. (2009)
Copper mine	NA	6.7	NA	NA	NA	65.5	0.1	Trace	Jordanov et al. (2009)
Copper mine	26.5	4.4	NA	218.0	NA	127.5	NA	0.4	Korac and Komberovic (2006)
Coal mine	NA	NA	NA	NA	NA	0.2	NA	4.8	Mishra et al. (2008)
Coal mine (AMD)	NA	2.6	NA	NA	NA	0.2	NA	112	Laus et al. (2007)
Industrial	23.0	7.4	NA	NA	NA	1.0	35	31.3	Kaczala et al. (2009)
Power plant	28.6	5.4	276.0	5.4	143.0	NA	Trace	NA	Junshum et al. (2007)
Industrial	NA	7.4	NA	NA	NA	0.7	NA	3.7	Rawat et al. (2003)
Hydrometallurgical	NA	5.4	468	NA	NA	0.02	42.4	NA	Kalderis et al. (2008)
Copper mine	23.2	6.7	1620	5	NA	50.8	NA	NA	Rozon-Ramilo et al. (2011)
Gold mine (PE)	32	9.0		1160		5.4	0.03	2.1	Benavente et al. (2011)
Gold mine (TPE)	31.6	7.7		991.0			0.06	0.3	Benavente et al. (2011)
Gold mine (PE)	33.2	9.8	4560	143000	2280	7.8	7.2	3.8	This study
Gold mine (TPE)	28.2	7.5	4890	75.2	2460	3.1	10.1	0.9	This study

NA = Not available; PE = process Effluent; TPE = Tailings Pond effluent; AMD = Acid Mine Drainage

PC5 shows a high positive loading value for As (0.80), moderate positive loading value for temperature (0.61) and high negative loading value for PO_4^{3-}, which is an indication of the geological source of contamination. It shows that temperature is an important physico-chemical parameter of the wastewater. PC6 on the other hand shows mixed sources of contamination with its high positive loading values for Cu (0.88) and SO_4^{2-} (0.74). The sources of sulphur (in SO_4^{2-}) and Cu are the arsenopyrite ore and the gold extraction process, respectively.

3.5 Conclusions

- The process effluent of the AngloGold Ashanti mine and the Sansu tailings dam wastewater quality was studied from June to September 2010. Almost all the parameters measured were found to exceed the permissible limit set by the Ghana EPA and the WHO guideline values. Copper, arsenic and cyanide were identified as the most toxic constituents. The wastewater from the Sansu tailings dam cannot be discharged into the environment without prior treatment. Any treatment technology for this effluent must focus on these contaminants. Sorption using agricultural and plant materials or natural sorbents could be a low-cost option for treating this effluent.

- The study showed that the Sansu tailings dam is an effective settler, removing 99.9% of TSS and 99.7% of the turbidity. Due to continuous loading of the tailings dam with fresh process effluent, the rate of biodegradation of cyanide and organic matter may be low. It is, however, able to contain the toxic effluent and prevent pollution of the environment with heavy metals and cyanide.

- Through the use of PCA, the sources of the pollutants in the wastewater have been identified as geological, process-related and mixed. It showed that the pollutants in the process and the tailings dam effluents have common sources, an indication that the Sansu tailings dam receives effluent from the gold extraction plant only.

3.6 Acknowledgements

The authors acknowledge funding from the Dutch Government under The Netherlands Fellowship Programme (NUFFIC award 32022513) and the Staff Development and Postgraduate Scholarship Scheme (Kumasi Polytechnic, Ghana). We are grateful to AngloGold Ashanti (Obuasi mine, Ghana) for providing facilities. We are particularly thankful to the manager of the environmental department, Mr. Owusu Yeboah for his guidance and support during the field studies.

3.7 References

Acheampong, M.A., Meulepas, R.J., Lens, P.N., 2010. Removal of heavy metals and cyanide fromgold mine wastewater. Journal of Chemical Technology & Biotechnology 85, 590-613.

Acheampong, M.A., Pereira, J.P.C., Meulepas, R.J.W., Lens, P.N.L., 2011. Biosorption of Cu(II) onto agricultural materials from tropical regions. Journal of Chemical Technology & Biotechnology 86, 1184-1194.

Acheampong, M.A., Pereira, J.P.C., Meulepas, R.J.W., Lens, P.N.L., 2011. Kinetics modelling of Cu(II) biosorption on to coconut shell and *Moringa oleifera* seeds from tropical regions. Environmental Technology 33, 409-417.

Ahonen, L., Tuovinen, O.H., 1992. Bacterial oxidation of sulfide minerals in column leaching experiments at suboptimal temperatures. Applied and Environmental Microbiology 58(2), 600-606.

American Public Health Association (1995) Standard methods for the examination of water and wastewater, 19th, ed., APHA, AWWWA, WEF, Washington, DC, USA, pp 1108

An, Q., Wu, Y., Wang, J., Li, Z., 2010. Assessment of dissolved heavy metal in the Yangtze River estuary and its adjacent sea, China. Environmental Monitoring and Assessment 164, 173-187.

Bai, J., Xiao, R., Cui, B., Zhang, K., Wang, Q., Liu, X., Gao, H., Huang, L., 2011. Assessment of heavy metal pollution in wetland soils from the young and old reclaimed regions in the Pearl River Estuary, South China. Environmental Pollution 159, 817-824.

Benavente, M., Moreno, L., Martinez, J., 2011. Sorption of heavy metals from gold mining wastewater using chitosan. Journal of the Taiwan Institute of Chemical Engineers 42, 976-988.

Boucabeille, C., Bories, A., Ollivier, P., Michel, G., 1994. Microbial degradation of metal complexed cyanides and thiocyanate from mining wastewaters. Environmental Pollution 84, 59-67.

Boyacioglu, H., 2006. Surface water assessment using factor analysis. Water SA 32 (3), 389-393.

Das, M., 2009. Identification of effluent quality indicators for use in irrigation - a factor analysis approach. Journal of Scientific & Industrial Research 68, 634-639.

Ghana EPA, 2007. Annual report: industrial effluent monitoring. Environmental Protection Agency, Accra, Ghana.

Ghana EPA, 2010. Environmental performance rating and disclosure: report on the performance of mining and manufacturing companies. Environmental Protection Agency, Accra, Ghana.

Gould, D.W., King, M., Mohapatra, B.R., Cameron, R.A., Kapoor, A., Koren, D.W., 2012. A critical review on destruction of thiocyanate in mining effluents. Minerals Engineering 34, 38-47.

Gupta, V.K., Gupta, M., Sharma, S., 2001. Process development for the removal of lead and chromium from aqueous solutions using red mud - an aluminium industry waste. Water Research 35, 1125-1134.

http://www.anglogold.com/subwebs/informationforinvestors/reports09/AnnualReport09/f/AGA_AR09.pdf, Accessed: 11/06/2012

Igbinosa, E.O., Okoh, A.I. 2009. Impact of discharge wastewater effluents on the physico-chemical qualities of a receiving watershed in a typical rural community. International Journal of Environmental Science and Technology 6 (2), 175-182.

Islam-ud-din, Khan, S., Hasham, A.E.-H., Ahmad, A., Houbo, S., Daqiang., C., 2010. Physico-chemical characteristics and bacterial diversity in copper mining wastewater based on 16S rRNA rene analysis. African Journal of Biotechnology 9(46), 7891-7899.

Jarboui, R., Chtourou, M., Azri, C., Gharsallah, N., Ammar, E., 2010. Time-dependent evolution of olive mill wastewater sludge organic and inorganic components and resident microbiota in multi-pond evaporation system. Bioresource Technology 101, 5749-5758.

Jordanov, H.S., Maletić, M., Dimitrov, A., Slavkov, D., Paunović, P., 2007. Waste waters from copper ores mining/flotation In: Bučbim' mine: characterization and remediation. Desalination 213, 65-71.

Junshum, P., Menasveta, P., Traichaiyaporn, .S., 2007. Water quality assessment in reservoirs and wastewater treatment system of the Mae Moh Power Plant, Thailand. Journal of Agriculture &Social Sciences 3(3), 1813-2235.

Kaczala, F., Marques, M., Hogland, W., 2009. Lead and vanadium removal from a real industrial wastewater by gravitational settling/sedimentation and sorption onto Pinus sylvestris sawdust. Bioresource Technology 100, 235-243.

Kalderis, D., Tsolaki, E., Antoniou, C., Diamadopoulos, E., 2008. Characterization and treatment of wastewater produced during the hydro-metallurgical extraction of germanium from fly ash. Desalination 230, 162-174.

Kyoseya, V., Todorova, E., Dombalov, I., 2009. Comparative assessment of the methods for destruction of cyanide used in gold mining industry. Journal of the University of Chemical Technology and Metallurgy 44(4), 403-408.

Laus, R., Geremias, R., Vasconcelos, H.L., Laranjeira, M.C.M., Fávere, V.T., 2007. Reduction of acidity and removal of metal ions from coal mining effluents using chitosan microspheres. Journal of Hazardous Materials 149, 471-474.

Li, M., Wu, Y.-J., Yu, Z.-L., Sheng, G.-P., Yu, H.-Q., 2007. Nitrogen removal from eutrophic water by floating-bed-grown water spinach (Ipomoea aquatica Forsk.) with ion implantation. Water Research 41, 3152-3158.

Li, M., Wu, Y.-J., Yu, Z.-L., Sheng, G.-P., Yu, H.-Q., 2009. Enhanced nitrogen and phosphorus removal from eutrophic lake water by Ipomoea aquatica with low-energy ion implantation. Water Research 43, 1247-1256.

Liu, W.X., Li, X.D., Shen, Z.G., Wang, D.C., Wai, O.W.H., Li, Y.S., 2003. Multivariate statistical study of heavy metal enrichment in sediments of the Pearl River Estuary. Environmental Pollution 121, 377-388.

Lucile, A., Lena, A., Björn, O., 2010. Geochemical evaluation of mine water quality in an open-pit site remediated by backfilling and sealing. In: Wolkersdorfer, C., Freund, A (eds): Mine Water & Innovative Thinking. p. 515-519, Sydney, Nova Scotia (CBU Press).

Lundgren, D.G., Silver, M., 1980. Ore leaching by bacteria. Annual Review of Microbiology 34, 263-283

Mahiroglu, A., Tarlan-Yel, E., Sevimli, M.F., 2009. Treatment of combined acid mine drainage (AMD)--Flotation circuit effluents from copper mine via Fenton's process. Journal of Hazardous Materials 166, 782-787.

Mantis, I., Voutsa, D., Samara, C., 2005. Assessment of the environmental hazard from municipal and industrial wastewater treatment sludge by employing chemical and biological methods. Ecotoxicology and Environmental Safety 62, 397-407.

Mazlum, N., Ozer, A., Mazlum, S. 1999. Interpretation of Water quality data by principal components analysis. Tr. J. of Engineering and Environmental Science 23:19-26

Micó, C., Recatalá, L., Peris, M., Sánchez, J., 2006. Assessing heavy metal sources in agricultural soils of an European Mediterranean area by multivariate analysis. Chemosphere 65, 863-872.

Mishra, V.K., Upadhyaya, A.R., Pandey, S.K., Tripathi, B.D., 2008. Heavy metal pollution induced due to coal mining effluent on surrounding aquatic ecosystem and its management through naturally occurring aquatic macrophytes. Bioresource Technology 99, 930-936.

Mitchell, J.W., 2009. An assessment of lead mine pollution using macro-invertebrates at Greenside Mines, Glenridding. Earth & E-environment 4, 27-57.

Mohan, D., Pittman Jr, C.U., Bricka, M., Smith, F., Yancey, B., Mohammad, J., Steele, P.H., Alexandre-Franco, M.F., Gómez-Serrano, V., Gong, H., 2007. Sorption of arsenic, cadmium, and lead by chars produced from fast pyrolysis of wood and bark during bio-oil production. Journal of Colloid and Interface Science 310, 57-73.

Monser, L., Adhoum, N., 2002. Modified activated carbon for the removal of copper, zinc, chromium and cyanide from wastewater. Separation and Purification Technology 26, 137-146.

Pena, E., Suárez, J., Sánchez-Tembleque, F., Jácome, A., Puertas, J., 2004. Characterization of polluted runoff in a granite mine, Galicia, Spain. In: Jarvis AP, Dudgeon BA, Younger PL (eds) Proceedings of the International Mine Water Association Symposium, University of Newcastle, Newcastle upon Tyne 1, 185-194.

Rawat, M., Moturi, M.C.Z., Subramanian, V., 2003. Inventory compilation and distribution of heavy metals in wastewater from small-scale industrial areas of Delhi, India. Journal of Environmental Monitoring 5, 906-912.

Rozon-Ramilo, L.D., Dubé, M.G., Rickwood, C.J., Niyogi, S., 2011. Examining the effects of metal mining mixtures on fathead minnow (Pimephales promelas) using field-based multi-trophic artificial streams. Ecotoxicology and Environmental Safety 74, 1536-1547.

Samudro, G., Mangkoedihardjo, S., 2010. Review on BOD, COD and BOD/COD ratio: a triangle zone for toxic, biodegradable and stable levels. International Journal of Academic Research, 2(4), 235-239.

Samudro, G.,, Mangkoedihardjo, S., 2012. Urgent need of wastewater treatment based on BOD footprint for aerobic conditions of receiving water. Journal of Applied Sciences Research 8(1), 454-457.

Sari, A., Tuzen, M., Citak, D., Soylak, M., 2007. Adsorption characteristics of Cu(II) and Pb(II) onto expanded perlite from aqueous solution. Journal of Hazardous Materials 148, 387-394.

Schneegurt, M.A., Jain, J.C., Menicucci, J.A., Brown, S.A., Kemner, K.M., Garofalo, D.F., Quallick, M.R., Neal, C.R., Kulpa, C.F., 2001. Biomass byproducts for the remediation of wastewaters contaminated with toxic metals. Environmental Science & Technology 35, 3786-3791.

Shukla, S.R., Pai, R.S., 2005. Adsorption of Cu(II), Ni(II) and Zn(II) on modified jute fibres. Bioresource Technology 96, 1430-1438.

Sisodia, R., Moundiotiya, C., 2006. Assessment of the water quality index of wetland Kalakho Lake, Rajasthan, Indian Journal of Environmental Hydrology 14, 1-11.

Smith, A., Struhsacker, D., 1987. Cyanide geochemistry in an abandoned heap leach system and regulations for detoxification. In: First International Conference on Gold Mining, Vancouver, BC, Canada, Chapter 29, 412-430.

USEPA, 2005. Treatment of Cyanide heap leaches and tailings: technical report. EPA 530-R-94-037, U. S. Environmental Protection Agency, Office of Solid Waste, Special Waste Branch, 401 M Street, SW, Washington, DC 20460.

USEPA, 2005. Metrics for expressing greenhouse gas emissions: carbon equivalents and carbon dioxide equivalents. EPA420-F-05-002. U. S. Environmental Protection Agency, Office of Transportation and Air Quality, 1200 Pennsylvania Ave., NW (6406J), Washington, DC 20460.

Vaugn, D.J., Craig, J.R., 1978. Mineral chemistry of metal sulphides. Cambridge University Press, Cambridge, ISBN 0521214890.

WHO, 1989. Health guidelines for use of wastewater in agriculture and aquaculture, technical report series 778, World Health Organisation, Geneva, Switzerland.

Yongming, H., Peixuan, D., Junji, C., Posmentier, E.S., 2006. Multivariate analysis of heavy metal contamination in urban dusts of Xi'an, Central China. Science of The Total Environment 355, 176-186.

Zagury, G.J., Oudjehani, K., Deschênes, L., 2004. Characterization and availability of cyanide in solid mine tailings from gold extraction plants. Science of The Total Environment 320, 211-224.

Zhang, Q., Li, Z., Zeng, G., Li, J., Fang, Y., Yuan, Q., Wang, Y., Ye, F., 2009. Assessment of surface water quality using multivariate statistical techniques in red soil hilly region: a case study of Xiangjiang watershed, China. Environmental Monitoring and Assessment 152, 123-131.

Chapter 4

4 Biosorption of Cu(II) onto agricultural materials from tropical regions

This chapter has been presented and published as:

Acheampong, M.A., Pereira, J.P.C., Meulepas, R.J.W., Lens, P.N.L., 2011. Biosorption of Cu(II) and As(III) from Gold Mining Wastewater using Agricultural Materials: Biosorbents Screening and Equilibrium Isotherm Studies. In: *Proceedings of the IWA International Conference on Water & Industry*, Valladolid, Spain (1 - 4 May 2011).

Acheampong, M.A., Pereira, J.P.C., Meulepas, R.J.W., Lens, P.N.L., 2011. Equilibrium Isotherm and Kinetics Studies of Cu(II) Biosorption from Gold Mining Wastewater by Agricultural Materials. In: *Proceedings of the 5th European Conference on Bioremediation*, Chania, Greece (4 -7 July 2011).

Acheampong, M.A., Pereira, J.P.C., Meulepas, R.J.W., Lens, P.N.L., 2011. Biosorption of Cu(II) onto agricultural materials from tropical regions. Journal of Chemical Technology & Biotechnology 86, 1184-1194.

Abstract

In Ghana, the discharge of untreated gold mine wastewater contaminates the aquatic systems with heavy metals such as copper (Cu), threatening ecosystem and human health. The undesirable effects of these pollutants can be avoided by treatment of the mining wastewater prior to discharge. In this work, the sorption properties of agricultural materials, namely coconut shell, coconut husk, sawdust and *Moringa oleifera* seeds for Cu(II) were investigated. The Freundlich isotherm model described the Cu(II) removal by coconut husk ($R^2 = 0.999$) and sawdust ($R^2 = 0.993$) very well and the Cu(II) removal by *Moringa oleifera* seeds ($R^2 = 0.960$) well. The model only reasonably described the Cu(II) removal by coconut shell ($R^2 = 0.932$). A maximum Cu(II) uptake of 53.9 mg g^{-1} was achieved using the coconut shell. The sorption of Cu(II) onto coconut shell followed the pseudo second-order kinetics ($R^2 = 0.997$). FTIR spectroscopy indicated the presence of functional groups in the biosorbents, some of which were involved in the sorption process. SEM-EDX analysis suggest an exchange of Mg(II) and K(I) for Cu(II) on *Moringa oleifera* seeds and K(I) for Cu(II) on coconut shell. This study shows that coconut shell can be an important low-cost biosorbent for Cu(II) removal. The results indicate that ion exchange, precipitation and electrostatic forces were involved in the Cu(II) removal by the biosorbents investigated.

Keywords: Biosorption, Removal, Waste-Water, Heavy Metals, Ion Exchange, Scanning Electron Microscopy (SEM)

4.1   Introduction

The emergent industrialization of developing countries is frequently related to serious environmental pollution. The pollution of water with toxic metals such as copper is dangerous due to their toxicity and non-biodegradability (Dundar et al., 2008; Gulnaz et al., 2005). Copper bearing mining wastes and acid mine drainages discharge significant quantities of dissolved copper and other heavy metals into the environment (Rahman and Islam, 2009; Bozic et al., 2009; Chojnacka, 2010; Kumar et al., 2006; Kumaressan et al., 2001; Kumari et al., 2006). The undesirable effects of heavy metal pollution can be avoided by treatment of the mining wastewater prior to discharge. In view of their toxicity and in order to meet regulatory discharge standards, it is essential to remove heavy metals from this wastewater before it is released into the environment. This helps to protect the environment and to guarantee the public health quality.

In Ghana, gold mining is a major economic activity and contributes significantly to the spillage and discharge of contaminated wastewater into rivers and other aquatic systems, threatening human life, flora and fauna (Acheampong et al., 2010). Although in 1999 the Environmental Assessment Regulation, Legislative Instrument (L.I.) 1652 was passed in Ghana to ensure that industrial discharges are within acceptable national standards, the Environmental Protection Agency (EPA) revealed that most of the industries were not respecting these standards (Ghana EPA, 2010). According to the EPA of Ghana, two main reasons caused this failure: the lack of adequate human resource capacity of industries to manage their wastewater and the unavailability of inexpensive wastewater treatment technologies in the country for these industries.

Some conventional methods are commonly used to remove heavy metal ions from mining wastewaters, namely filtration, membrane filtration, ion-exchange, reverse osmosis, chemical precipitation, coagulation, evaporation recovery, and electrochemical technologies (Ayoub et al., 2001). However, they often require high energy consumption, and generate large amounts of waste products that need further treatment and disposal. Therefore, most of these methods are expensive when treating large amounts of wastewater containing low concentrations of heavy metals, and thus cannot be used at large scale (Wang and Chan, 2009).

In order to provide an alternative low cost solution to diffuse pollution problems, several biological technologies have been investigated for the removal of pollutants from wastewaters. Biosorption emerged as the most efficient and cost effective method for the removal of toxic metals from contaminated wastewaters, and thus received particular attention in recent years (Gadd, 2009). Biosorption is based on the ability of biological materials to remove heavy metals from solutions due to their metal binding capacities, by means of mechanisms such as ion exchange, electrostatic force and precipitation. All biological materials have a natural affinity for metal binding; therefore biosorption allows the use of low-cost and locally available materials (Arief et al., 2008; Ulmanu et al., 2003; Amuda et al., 2007).

Agricultural materials and industrial by-products such as coconut materials, *Moringa oleifera* seed powder, and sawdust have been suggested as promising adsorbents for the removal of various heavy metals, such as copper and arsenic (Acemioglu et al., 2004; Pino et al., 2006; Habib et al., 2007; Sajidu et al., 2008). These materials are readily available in Ghana, and therefore could provide a potential low cost solution for the treatment of gold mine wastewater. The goal of this work is to investigate the sorption capacity, affinity and kinetics of coconut shell, coconut husk, *Moringa oleifera* seeds and sawdust for Cu(II). The potential of these biomaterials as low cost biosorbents was studied by means of batch sorption tests.

4.2   Materials and methods

4.2.1   Sorption materials
The coconut shell, coconut husk, sawdust and *Moringa oleifera* seeds were obtained from Kumasi (Ghana). The materials were washed with distilled water and dried at 105^0C for 24 h. The coconut shells were grinded using a Peppink hammer mill and the coconut husk was grinded with a Retsch knife mill. A laboratory blender was used to grind the *Moringa oleifera* seeds, while the sawdust was received in a grinded form. Subsequently, the materials were sieved into fractions with the following particle size ranges: 0.25 - 0.5 mm, 0.5 - 0.8 mm, 0.8 - 1.0 mm, 1.0 - 1.4 mm and 1.4 - 1.6 mm. The four biosorbents used in this study have very rough and corrugated surfaces.

4.2.2   Metal ion solutions
Batch sorption experiments were done with copper chloride (CuCl$_2$) solution with Cu(II) concentrations of 5, 10, 15, 20, 25 and 50 mg L^{-1}. These concentrations were selected to cover the concentration range of the actual gold mine wastewater (Acheampong et al., 2010) and to enable the derivation of sorption isotherms. For coconut shell, 100 and 200 mg L^{-1} Cu(II) solutions were used as well, because it has the capacity to remove Cu(II) at these concentrations. All chemicals used were of analytical grade.

4.2.3   Biosorption experiments

Sorption experiments were conducted in triplicates in 250 mL Erlenmeyer flasks filled with 100 mL metal solution and 2.0 g (dry weight) of sorbent material of a 0.5 - 0.8 mm particle size range. Immediately after adding the sorbent material, the pH was adjusted to 7.0 ± 0.2 by adding HCl and NaOH solutions. The flasks were agitated for 24 hours at 30 ± 0.2 ^{0}C on an orbital shaker at 100 rpm. After 24 hours, the solutions were filtered with 0.45 μm filter paper (Schleider & Schuell, No. 595 1/2). The filtrate was collected and diluted with deionised water and acidified with 65% HNO_3 for copper ion determination by atomic absorption spectrometry (AAS). In order to assess the copper uptake by the 4 sorbents, samples of the sorbent materials, before and after sorption, were digested with concentrated nitric acid (65%) in a microwave (MARS 5, CEM Corporation) according to standard methods (APHA, 1995). Subsequently, the dissolved copper was quantified by AAS.

Blank tests (without sorbent) were performed in duplicate to assess the occurrence of precipitation at the experimental conditions. The effect of sorbent concentration was investigated by varying the amount of coconut shell (0.5 - 0.8 mm) added to the 100 ml of 10 mg L^{-1} Cu(II) solution from 1.0 to 15.0 g. Besides varying the metal ion and sorbent concentrations, a batch sorption experiment with 2.0 g coconut shell and 10 mg L^{-1} Cu(II) solution was done to determine the effect of the particle size, by comparing the 0.25 - 0.5 to 1.4 - 1.6 mm ranges.

The kinetics of the copper removal by coconut shell were studied by frequent (every 30 min) sampling over the course of 10.5 h experiments in 5 replicate bottles. The residual copper concentrations of each bottle was then analysed using AAS.

4.2.4   Characterisation of the biosorbents

4.2.4.1   *Density*
Density measurements were done in a Quantachrome Ultrapycnometer 1000 (Tsai et al., 2006). Prior to the actual analysis; the samples were pre-treated at 105^0C for 10 hours in order to remove possible moisture and other volatile compounds. The weight loss obtained by the pre-treatment was recorded as well. The measurements were carried out according to the ASTM C 604 standard test method (Gregg and Sing, 1982).

4.2.4.2   *Surface area (S_{BET})*
The multi-point BET surface area (Tai et al., 2006) has been determined in the relative pressure range 0.05-0.25 (Gregg et al., 1982) for the four biosorbents. As a relatively low surface area was expected, a maximum sample weight has been used in the analysis. Part of the adsorption isotherm with N_2 as adsorptive was recorded at 77 K on a Micromeritics TriStar 3000. Prior to the adsorption measurement, the samples were degassed under vacuum at 105^0C for 16 hours to remove volatile impurities and moisture which may have been adsorbed on the surface previously (Singh and Pant, 2004).

4.2.4.3   *Porosity*
The biosorbents were pre-treated at 105^0C in vacuum for 10 hours. The dried samples were then transferred and weighed in the sample holder. The measurements were carried out on a CE Instruments Pascal 140 (low pressure) and a CE Instruments

Pascal 440 (high pressure) (Gregg et al., 1982). The applied pressure ranged from vacuum to 400 kPa for the Pascal 140 measurement, which covered pore diameters from approximately 100 to 4 μm, and 0.1 MPa to 400 MPa for the Pascal 440, which covered pore diameters down to 4 nm.

4.2.4.4   Acid titrations for Point of Zero Charge (PZC) determination
Potentiometric mass titration (PMT) was used to determine the Point of Zero Charge (PZC) of the four biosorbents as described by Faria et al. (1998), and Fiol and Villaescusa (2009). All experiments were performed at $30 \pm 0.2\ ^0C$ under a N_2 atmosphere. A suspension of three different masses of sorbent materials in the range of 5 to 30 g were put into contact with 200 mL of the 0.03 M KNO_3 and the mixture was agitated for 24 hours at a speed of 150 rpm. The suspension was then titrated by drop-wise (0.05 mL) addition of 0.1 M HNO_3 under continuous stirring (150 rpm). The pH of the solution was continuously recorded.

4.2.4.5   Fourier transform infrared spectroscopy
Fourier transform infrared (FTIR) spectra of air-dried samples of the sorbent materials were recorded on KBr pellets at room temperature using a *Nicolet* 380 FTIR spectrometer (Thermo scientific, Waltham, MA-US) equipped with a DTGS detector. The sample compartment was flushed with dry air to reduce interference of H_2O and CO_2. The data point resolution of the spectra was 2 cm^{-1} and 25 scans were accumulated for each spectrum. Data analysis focused on the 400 - 4000 cm^{-1} region.

4.2.4.6   Scanning Electron Microscopy and X-ray microanalysis
Scanning Electron Microscopy (SEM) analysis was performed on a Hitachi TM-1000 at magnifications of 10 to 10,000 to elucidate the porous properties of the biosorbents. X-ray microanalysis was performed to detect the atomic composition of the biosorbents using a Quantax energy dispersive X-ray spectrometry system (EDX).

4.2.5   Analytical techniques
The concentration of copper in the filtrate was measured with an AAS (Perkin Elmer, model AAnalist200), equipped with an air-acetylene flame. The minimum and the maximum detection limits of the AAS for Cu(II) are 0.01 mg L^{-1} and 1.5 mg L^{-1}, respectively. As a quality assurance procedure, a 3-point calibration through zero was applied with blank as well as standard solutions of 0.5, 1.0 and 1.5 mg L^{-1}. The resulting linear calibration line ($r^2 \geq 0.996$) was the requirement for any sample measurement. Furthermore, after every 10 measurements, the 1.0 mg L^{-1} standard solution was checked for accuracy. The calibration was repeated in cases where incorrect values were obtained. Finally, the standard solution (1.3 mg L^{-1}) for the required absorbance (0.20) was checked as part of the quality assurance process. The pH was measured with a SenTix21 pH electrode (WTW model pH323). The instrument was calibrated using buffer solutions with pH values of 4.00, 7.00 and 10.00.

4.2.6   Calculations
The metal uptake per gram of sorbent and the percentage removal were calculated according to the following equations:

$$q_e = \frac{(C_0 - C_e)}{m} \times V \qquad\qquad\qquad (4\text{-}1)$$

$$\mathrm{Re}\,moval(\%) = \frac{(C_0 - C_e)}{C_0} \times 100 \qquad (4\text{-}2)$$

Where q_e is the equilibrium adsorption capacity (mg g^{-1}), C_0 is the initial concentration (mg L^{-1}) of metal ions in solution, C_e is the equilibrium concentration (mg L^{-1}) of metal ions in solution, V is the volume of aqueous solution (L) and m is the dry weight of the adsorbent (g).

4.2.7   Sorption isotherms

4.2.7.1   The Langmuir isotherm model

The Langmuir isotherm model is valid for monolayer adsorption onto surface containing a finite number of identical sorption sites and it is represented by the following equation (Langmuir, 1916):

$$q_e = \frac{q_e K_L C_e}{1 + K_L C_e} \qquad (4\text{-}3)$$

where q_e is the amount of metal adsorbed per specific amount of adsorbent (mg g^{-1}), C_e is the equilibrium concentration of the solution (mg L^{-1}) and q_m is the maximum amount of metal ions required to form a monolayer (mg g^{-1}). The Langmuir equation can be rearranged to a linear form for the convenience of plotting and determination of the Langmuir constant (K_L) as below. The values of q_m and K_L can be determined from the linear plot of C_e/q_e versus C_e:

$$\frac{C_e}{q_e} = \frac{1}{K_L q_e} + \frac{1}{q_m} C_e \qquad (4\text{-}4)$$

The essential characteristics of the Langmuir isotherm parameters can be used to predict the affinity between the sorbate and the sorbent using the separation factor or dimensionless equilibrium parameter "R_L", expressed as in the following equation (Rahman and Islam, 2009):

$$R_L = \frac{1}{1 + K_L C_0} \qquad (4\text{-}5)$$

where K_L is the Langmuir constant and C_0 is the initial concentration of the metal ion. The value of the separation factor R_L provides important information about the nature of adsorption. The value of R_L is between 0 and 1 for favourable adsorption, while $R_L > 1$ represents unfavourable adsorption and $R_L = 1$ represents linear adsorption. The adsorption process is irreversible if $R_L = 0$

4.2.7.2   The Freundlich isotherm model

The Freundlich equation is purely empirical based on sorption on a heterogeneous surface. It is commonly described by the following equation:

$$q_e = K_F C_e^{1/n} \qquad (4\text{-}6)$$

In its linear form, this equation assumes the following form (Rahman and Islam, 2009):

$$\log q_e = \log K_F + (1/n)\log C_e \qquad (4\text{-}7)$$

where q_e is the metal uptake (mg g^{-1}) at equilibrium, K_F is the measure of the sorption capacity, $1/n$ is the sorption intensity, and C_e is the final Cu(II) concentration in solution, or equilibrium concentration (mg L^{-1}). The Freundlich isotherm constants K_F and $1/n$ are evaluated from the intercept and the slope respectively, of the linear plot of log q_e versus log C_e.

4.2.7.3   *Dubinin-Radushkevich isotherm (D-R) model*
This isotherm assumes that the characteristics of the sorption curve are related to the porosity of the biosorbents (Abdelwahab, 2007). This model was chosen to evaluate the mean energy of sorption. It is represented in the linear form by the equation (Igwe and Abia, 2007):

$$\ln q_e = \ln Q_D - 2B_D RT \ln\left(1 + 1/C_e\right) \qquad (4\text{-}8)$$

where Q_D is the theoretical maximum capacity (mol g^{-1}), B_D is the D-R model constant (kJ mol^{-1} K^{-1}), T is the absolute temperature (K) and R is the gas constant (kJ mol^{-1}). The value of Q_D and B_D can be obtained from the intercept and slope of the plot of ln (q_e) versus ln (1 + 1/C_e). The mean energy of sorption, E (kJ mol^{-1}) is calculated from the relation (Igwe and Abia, 2007; Horsfall Jr. et al., 2004):

$$E = 1\sqrt{2B_D} \qquad (4\text{-}9)$$

4.2.7.4   *Flory-Huggins isotherm model*
Flory-Huggins model was chosen in order to estimate the number of Cu(II) occupying sorption sites. The generalised and linearised forms of the model are given by equations (4-10) and (4-11), respectively (Mahamadi and Nharingo, 2010):

$$C_0 = \frac{\theta}{K_{FH}(1-\theta)^n} \qquad (4\text{-}10)$$

$$\log \frac{\theta}{C_0} = \log K_{FH} + n\log(1-\theta) \qquad (4\text{-}11)$$

where $\theta = (1 - C_e/C_0)$ is the degree of coverage, n is the number of metal ions occupying sorption sites and K_{FH} is the equilibrium constant of adsorption. From equation (11), a plot of log (θ/C_0) versus log (1 - θ) gives a straight line from which n and K_{FH} can be calculated from the slope and intercept. The Gibbs free energy of adsorption is computed from the equation below (Mahamadi and Nharingo, 2010):

$$\Delta G = -RT \ln K_{FH} \qquad (4\text{-}12)$$

4.2.7.5   Temkin isotherm model

The Temkin isotherm model assumes that the heat of sorption decreases with the coverage as a result of adsorbate-adsorbent interaction (Vijayaraghavan et al., 2009). The linearised form is expressed by the following equation (Mahamadi and Nharingo, 2010):

$$q_e = \frac{RT}{b_T} \ln K_T + \frac{RT}{b_T} \ln C_e \qquad\qquad (4\text{-}13)$$

where K_T (L g^{-1}) is the Temkin isotherm constant, b_T (J mol^{-1}) is a constant related to the heat of sorption and R (8.314 J mol^{-1} K^{-1}) is the gas constant. A plot of q_e versus ln (C_e) gives a straight line from which K_T and b_T can be evaluated from the slope and the intercept.

4.2.8   Sorption kinetics
4.2.8.1   Pseudo first-order kinetic model

The linearised form of the pseudo first-order kinetic model (Abdelwahab, 2007) was applied to the experimental data to describe the kinetics of Cu(II) sorption. This equation is given below:

$$\log(q_e - q_t) = \log(q_e) - \frac{k_1}{2.303} t \qquad\qquad (4\text{-}14)$$

where q_t and q_e are the amount of solute sorbed per mass of sorbent (mg g^{-1}) at any time and equilibrium, respectively, and k_1 is the rate constant of first-order sorption (min^{-1}). The straight-line plot of log (q_e - q_t) against t gives log (q_e) as slope and intercept equal to $k_1/2.303$. Hence the amount of solute sorbed per gram of sorbent at equilibrium (q_e) and the first-order sorption rate constant (k_1) can be evaluated from the slope and the intercept.

4.2.8.2   Pseudo second-order kinetic model

The Lagergren pseudo second-order kinetic model was applied to the experimental data to describe the kinetics of Cu(II) sorption. The model is represented by Eq. 15 (Ho and McKay, 1999):

$$\frac{t}{q_t} = \frac{1}{k_2 q_e^{\,2}} + \left(\frac{1}{q_e}\right) t \qquad\qquad (4\text{-}15)$$

The initial sorption rate is defined by the following equation (Kalavathy et al., 2005):

$$h = k_2 q_e^{\,2} \qquad\qquad (4\text{-}16)$$

where k_2 is the rate constant, q_t is the metal uptake capacity at any time t.

4.3   Results

4.3.1   Sorption isotherms of Cu(II)

The sorption isotherms of Cu(II) are shown in Figs. 4-1 and 4-2. The experimental data were fitted with the Langmuir, Freundlich, Temkin, Dubinin-Radushkevich and Flory-Huggins isotherm models and the results presented in Table 4-1.

The coconut shell isotherms show an exponential increase of the sorption capacity (q) with increasing equilibrium concentration. The exponential relation between C_{eq} and q is observed both at the 0 - 1.5 mg L^{-1} range (Figure 4-2, obtained within sorbent dosage experiment) and the 0 - 10 mg L^{-1} range (Figure 4-1A, obtained in the experiment with different initial concentrations). These results do not fit well with any of the models ($R^2 \leq 0.932$) (Table 4-1), whereas the Cu(II) removal by coconut husk ($R^2 = 0.999$) and sawdust ($R^2 = 0.993$) was well described by the Freundlich isotherm model (Table 4-1). The fits with the Langmuir, Temkin, Dubinin-Radushkevich and Flory-Huggins isotherm models were comparatively less good. The best models describing Cu(II) sorption by *Moringa oleifera* seeds were the Dubinin-Radushkevich isotherm model ($R^2 = 0.989$) and the Freundlich isotherm model ($R^2 = 0.961$).

Table 4-2 shows the initial and final amount of Cu in the liquid and solid phases during sorption experiments with the four studied materials. The gab in the Cu balance for coconut shell, coconut husk and the *Moringa oleifera* seeds was less than 3%.

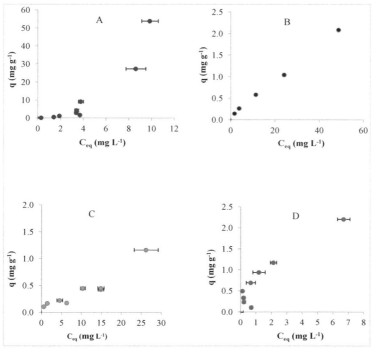

Figure 4-1: Isotherms for Cu(II) sorption onto coconut shell (A), coconut husk (B), saw dust (C) and *Moring oleifera* seeds (D), obtained by varying the initial Cu(II) concentration. Experimental conditions: sorbent concentration = 20 g L^{-1}, particle size = 0.5 - 0.8 mm, contact time = 24 h, mixing rate = 100 rpm, temperature = 30 ± 1 ^{0}C, pH = 7.0 ± 0.2

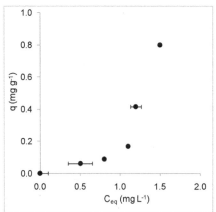

Figure 4-2: Isotherm for Cu(II) sorption onto coconut shell, obtained by varying the sorbent concentration from 10 – 150 g L^{-1}. Experimental conditions: initial Cu(II) concentration = 10 mg L^{-1}, particle size = 0.5 - 0.8 mm, contact time = 24 h, mixing rate = 100 rpm, temperature = 30 ± 0.2 ^{0}C, pH = 7.0 ± 0.2

Table 4-1: Coefficients of five different sorption isotherm models for Cu(II) removal by coconut shell, coconut husk, sawdust and *Moringa oleifera* seeds and the coefficient of determination (R^2). Experimental conditions: particle size = 0.5 - 0.8 mm, contact time = 24 h, mixing rate = 100 rpm, temperature = 30 ± 0.2 ^{0}C, pH = 7.0 ± 0.2

Isotherm parameters	Biosorbents			
	Coconut shell	Coconut husk	Sawdust	*Moringa oleifera* seeds
Langmuir				
q$_{max}$	4.21	4.62	179	238
K$_L$	0.104	0.015	0.933	4.02
R$_L$	0.490	0.870	0.097	0.024
R^2	0.617	0.719	0.954	0.160
Freundlich				
1/n	0.451	0.0407	0.0409	0.264
K$_F$	1.08	1.23	1.17	3.10
R^2	0.932	0.999	0.993	0.961
Temkin				
A	2.03	3.16	1.19	8.348
b	316	6.78 x10^3	1.14 x10^4	4.90 x10^3
B	7.17	0.335	0.200	0.463
R^2	0.354	0.674	0.610	0.930
Dubinin-Radushkevich				
Q$_D$	14.1	2.95	2.15	2.32
B$_D$	3.74x10^{-7}	1.32 x10^{-7}	1.54 x10^{-7}	1.32 x10^{-7}
E	3.06x10^{-4}	1.82 x10^{-4}	1.96 x10^{-4}	1.82 x10^{-4}
R^2	0.703	0.978	0.571	0.989
Flory-Huggins				
n	1.81	9.04	2.58	1.38
K$_{FH}$	4.24	4.36 x10^3	381	24.1
ΔG^0_{ads}	-3.28	-2.43	-1.35	-7.23
R^2	0.874	0.957	0.771	0.465

Table 4-2: Copper balances over the 24 h sorption experiments with coconut shell, coconut husk, saw dust and *Moringa oleifera* seeds (N=3). Experimental conditions: sorbent concentration = 20 g L^{-1}, particle size = 0.5 - 0.8 mm, mixing rate = 100 rpm, temperature = 30 ± 0.2 $^{\circ}$C, pH = 7.0 ± 0.2

Sorbent	Initial amount of Cu in solution (mg)	Initial amount of Cu in sorbent (mg)	Final amount of Cu in solution (mg)	Final amount of Cu in sorbent (mg)	Recovery (%)
Coconut shell	2.44 (±0.03)	0.03 (±0.00)	0.30 (±0.00)	2.10 (±0.02)	97
Coconut husk	0.93 (±0.01)	0.01 (±0.00)	0.38 (±0.00)	0.56 (±0.00)	100
Saw dust	1.01 (±0.01)	0.01 (±0.00)	0.61 (±0.01)	0.56 (±0.00)	115
Moringa seeds	1.00 (±0.00)	0.01 (±0.00)	0.01 (±0.00)	1.00 (±0.01)	100

4.3.2   Effect of sorbent particle size

Table 4-3 presents the equilibrium concentration and removal percentage in sorption experiments with a 20 mg L^{-1} Cu(II) solution and 20 g L^{-1} coconut shell with different particle sizes. The removal percentage decreased from the smaller to the larger particle size range due to decreased surface area. However, the removal was only 7.4% better when the particle diameter was decreased from 1.4 - 1.6 mm to 0.25 - 0.5 mm.

4.3.3   Kinetics of Cu(II) sorption by coconut shell

Figure 4-3A shows the removal of Cu(II) in time by 20 g L^{-1} coconut shell. Figures 4-3B and 4-3C show the linear plot of log (qe - qt) versus t for the Lagergren pseudo first-order model and t/q$_t$ versus t for the Lagergren pseudo second-order model for the biosorption of Cu(II). The correlation coefficient (R^2) and the equilibrium rate constant of the pseudo first-order sorption (k$_1$) are 0.819 and 2.022 × 10^{-4} min^{-1}, respectively. For the second-order model, the correlation coefficient (R^2) and the rate constant (k$_2$) are 0.997 and 4.785 × 10^{-2} g mg^{-1} min^{-1}, respectively.

4.3.4   Physical characterization of the biosorbents

The coconut husk and sawdust have a similar pore volume (Table 4-4) and pore size distribution of the interstitial space. The specific volume and porosity of coconut shell and *Moringa oleifera* seeds are lower than those of the other two materials (Table 4-4).

Table 4-3: Equilibrium concentrations (C$_{eq}$) and removal percentages of Cu(II) after sorption onto coconut shell with different particle size ranges. Experimental conditions: initial Cu(II) concentration = 10 mg L^{-1}, sorbent concentration = 20 g L^{-1}, contact time = 24 h, mixing rate = 100 rpm, temperature = 30 ± 1 $^{\circ}$C, pH = 7.0 ± 0.2

Particle size (mm)	C$_{eq}$ (mg L^{-1})	Removal (%)
0.25 - 0.5	0.6 (±0.02)	93.60 (±0.02)
0.5 - 0.8	0.9 (±0.00)	90.77 (±0.00)
0.8 - 1.0	1.0 (±0.01)	89.90 (±0.01)
1.0 - 1.4	1.0 (±0.01)	89.90 (±0.01)
1.4 - 1.6	1.4 (±0.02)	86.20 (±0.02)

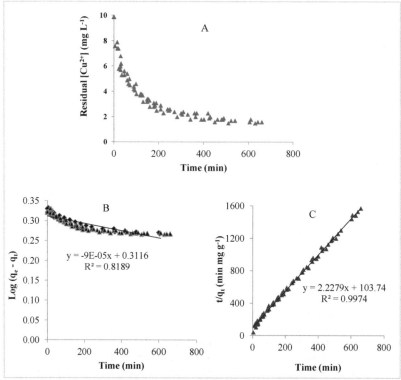

Figure 4-3: Cu(II) sorption onto coconut shell in time (A) and the determination of the pseudo first-order (B) and second-order (C) rate constant. Experimental conditions: initial Cu(II) concentration = 10 mg L^{-1}, sorbent concentration = 20 g L^{-1}, particle size = 0.5 - 0.8 mm, mixing rate = 100 rpm, temperature = 30 ± 1 ^{0}C, pH = 7.0 ± 0.2.

4.3.5   Acid titrations for Point of Zero Charge determination

The point of zero charge (PZC) is an adsorption phenomenon which describes the condition when the electrical charge density on a surface is zero (Faria et al., 1998). Figure 4-4 shows the potentiometric curves for the four biosorbents investigated. The common intersection point of the titration curves with the blank is the pH at PZC (pH$_{PZC}$). From the curves (Figure 4- 4), the pH$_{PZC}$ for coconut shell, coconut husk, sawdust and *Moringa oleifera* seeds were identified as 6.5, 5.3, 7.2 and 6.9, respectively. The titration curve of *Moringa oleifera* seeds is flatter than those of the other materials. This indicates a larger capacity to take up protons (buffering capacity). Therefore, the capacity to take up cationic metals by ionic exchange is probably also bigger. The titration curve of coconut shell is the steepest, indicating a low pH buffering capacity.

Table 4-4: Physical characteristics of the biosorbents

Biosorbent	Specific volume* (cm^3 g^{-1})	Porosity (%)	Density (g cm^{-3})	Specific surface area (m^2 g^{-1})
Moringa oleifera seeds	0.46	34	1.14	0.1
Coconut shell	0.90	55	1.35	0.4
Coconut husk	1.78	72	1.46	0.9
Sawdust	1.77	77	1.63	0.7

*Including the pore volume

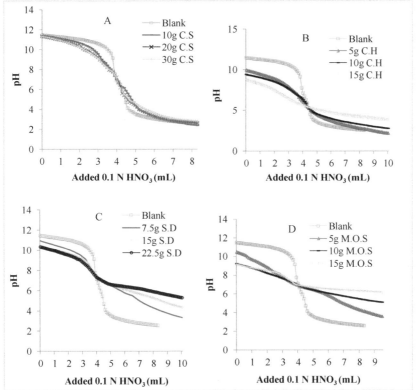

Figure 4-4: Potentiometric titration curves for coconut shell (A), coconut husk (B), saw dust (C) and *Moringa oleifera* seeds (D).

4.3.6   Fourier transform infrared spectroscopy

Figures 4-5A, 4-5B and 5C show that coconut shell, coconut husk and sawdust have identical FTIR spectra, both in shape and intensities. The peaks at around 3385, 2904, 2427, 1735, 1647 and 1596, 1384, 1241 and 115-1056 cm^{-1} represent hydroxyl groups, C–H stretching vibrations, carboxyl acid groups, C=O stretching vibrations, carboxyl groups, aromatic CH and carboxyl groups, carbonate structures, C-O stretching vibrations and C–OH stretching vibrations, respectively. Carboxylic acids showed a broad, intensely sharp -OH stretching absorption from 3650 to 3590 cm^{-1}, although the bands were dominated by the -OH stretch due to bound water. The presence of amine groups was also possible. The broad peak at 3400–3000 cm^{-1} describes N-H stretches. 1029 cm^{-1} is the stretching vibration of C–O–C and O–H of polysaccharides. 582 cm^{-1} shows the characteristic band of –SH$_2$ and –PO$_4$ functional groups. The band at around 1420 cm^{-1} is assigned to symmetric COO–. The band at 2854.2 cm^{-1} and 2925.2 cm^{-1} represents –NH$_3$$^+$ symmetric stretching. This suggests that coconut shell, coconut husk and sawdust are dominated by -OH groups; in addition they contain other carbon-oxygen compounds (Figure 4-5). The *Moringa oleifera* seeds on the other hand, are dominated by carboxyl groups and nitrogen-hydrogen compounds such as -NH$_3$$^+$ (Figure 4-5D).

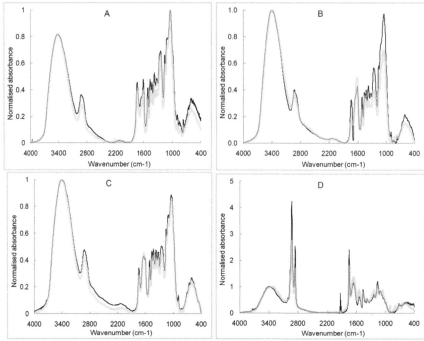

Figure 4-5: FTIR spectra of coconut shell (A), coconut husk (B), saw dust (C) and *Moring oleifera* seeds (D) before (grey) and after Cu(II) sorption (black). Experimental conditions: initial Cu(II) concentration = 10 mg L^{-1}, sorbent concentration = 20 g L^{-1}, particle size = 0.5 - 0.8 mm, contact time = 24 h, mixing rate = 100 rpm, temperature = 30 ± 0.2 ^{0}C, pH =7.0 ± 0.2.

For the coconut shell, a decrease in intensity of the band at 2094 cm^{-1} (C-H stretching vibrations) was observed after Cu(II) sorption. In addition, there was a shift in the band at 1647 - 1735 cm^{-1} (aromatic CH and C=O stretching vibrations). In the case of the coconut husk, the peak at 1735 cm^{-1} (C=O stretching vibrations) disappeared after Cu(II) sorption. At the same time, there was an increase in the intensity of the peak at 1647 cm^{-1} (aromatic CH), while the peak at 1029 cm^{-1} (vibrations of C-O-C and O-H of polysaccharides) decreased. A shift in the band at 1735 - 2904 cm^{-1} (C=O stretching vibrations, C-H stretching vibrations and carboxylic acid groups) was observed for the sawdust after Cu(II) sorption. For the *Moringa oleifera* seeds, a shift of the band at 3200 - 3400 cm^{-1} (N-H stretching vibrations) and an increase in the intensity of the peak at 1735 cm^{-1} (C=O stretching vibrations) was observed after Cu(II) sorption.

4.3.7   Scanning Electron Microscopy and X-ray microanalysis analyses

Figure 4-6: Scanning electron micrographs (180 times of magnification) of coconut shell (A) and *Moringa oleifera* seeds (B).

Table 4-5 shows the normalized elemental composition of coconut shell and *Moringa oleifera* seeds obtained with X-ray microanalysis. Light elements like H, N and O are not measured with EDX (Gregg and Sing, 1982). Figure 4-6 shows scanning electron microscopy pictures (180 times of magnification) of coconut shell and *Moringa oleifera* seeds.

The SEM-EDX results (Table 4-5) indicate an increase in the copper content in the biomass after sorption. In coconut shell, copper is the most abundant element. The disappearance of the 1.4 keV and 3.4 keV peaks, corresponding to Mg(II) and K(I), and the increase of the 8 keV and 8.8 keV peaks, corresponding to Cu(II), after copper sorption by *Moringa oleifera* seeds indicate ion exchange took place. For coconut shell, this was only observed for K(I).

Table 4-5: Normalized element composition of *Moringa oleifera* seeds and coconut shell, obtained with Energy-dispersive X-ray spectroscopy, before (blank) and after sorption experiments.

Biomass Type	Initial Cu conc. (mg L⁻¹)	Magnification	Elemental composition (%)						
			Mg	Si	P	S	K	Ca	Cu
MOS	blank	1200	2.7	0	17	55	24	0	1.4
MOS	blank	5000	3.3	0	19	58	20	0	0
MOS	blank	6000	2.8	14	22	46	15	0	0.4
MOS	5.0	180	0	0	44	53	0	0	3.2
MOS	5.0	1200	0	0	38	60	0	0	1.9
MOS	5.0	5000	3.5	4.9	37	45	0	8.5	0.9
CS	blank	800	0	0	0	0	71	0	29.0
CS	5.0	800	0	0	0	0	0	0	100
CS	50	800	0	0	0	0	0	0	100

MOS = *Moringa oleifera* seeds; CS = coconut shell

4.4   Discussion

4.4.1   Physical characterization of the biosorbents

The four biosorbents display a low specific surface area (Table 4-4), suggesting the absence of distinct porosity by micro- or mesopores. Despite this, a significant difference exists between the various biosorbents studied. The specific surface area (0.9 m² g⁻¹ and 0.7 m² g⁻¹) of coconut husk and saw dust, respectively, were the highest of the four materials tested. The specific surface area of coconut shell, which had the highest sorption capacity, was lower (0.4 m² g⁻¹). Therefore, specific surface area alone cannot explain the differences in sorption capacity. The specific surface area of *Moringa oleifera* seeds was the lowest ($0.1 \ m^2 \ g^{-1}$).

4.4.2   Sorption isotherms

The biosorption process in this study was well described by the linear form of the Freundlich equilibrium isotherm model which yielded R^2 values of 0.9320, 0.9985, 0.9927 and 0.9606 for the biosorption of Cu(II) onto coconut shell, coconut husk, sawdust and *Moringa oleifera* seeds, respectively. The adsorption intensity $1/n$ was less than 1 for Cu(II) sorption onto all biosorbents investigated (Table 4-1). This indicates favourable sorption (Rahman and Islam, 2008).

Even though the Langmuir isotherm model described the sorption process poorly, the separation factor for all four biosorbents was found to be between 0 and 1 (Table 4-1), indicating a favourable biosorption process (Rahman and Islam, 2008). The Langmuir model contains a number of assumptions which include that (a) all binding sites possess an equal affinity for the adsorbate, (b) adsorption is limited to the formation of a monolayer and (c) the number of adsorbed species does not exceed the total number of surface sites, that is, there is a 1:1 stoichiometry between surface adsorption sites and adsorbate (Gadd, 2009). It is likely that none of these assumptions apply in biological systems. Furthermore, the Langmuir isotherm assumes a finite number of uniform adsorption sites and the absence of lateral interactions between adsorbed species. These assumptions are clearly invalid for most complex systems including biological materials (Gadd, 2009). As a result, the Langmuir model could not fit the data well as compared to the Freundlich model which takes surface roughness into account.

The high Q_D value obtained from the D-R model (Table 4-1) for coconut shell indicates a high sorption capacity compared to the other biosorbents studied. The very low value of the mean sorption energy (Table 4-1) shows that the forces of attraction are Van der Waal and not chemical bonding. Therefore, the sorption process observed in this study is physical adsorption (i.e. physisorption). In general, $\Delta G^0_{ad} \leq -20$ kJ mol^{-1} represents physisorption, while in chemisorption $\Delta G^0_{ad} \geq -40$ kJ mol^{-1} (Horsfall Jr. et al., 2004; Ayawei et al., 2005). Furthermore, the negative value of ΔG^0_{ad} (Table 4-1) confirmed the feasibility of the process and the spontaneous nature of the sorption. The ΔG^0_{ad} is less than -20 kJ mol^{-1} for all the biosorbents studied, indicating a physisorption process and electrostatic interaction between Cu(II) and the biosorbents.

The low value of the Temkin isotherm constant b_T obtained (Table 1) indicates a weak interaction between the sorbents and Cu(II), supporting an ion exchange mechanism. Ho et al. (1995) reported that the typical range of the binding energy for the ion exchange mechanism is 8 to 9 kJ mol^{-1}.

The performance of the coconut husk and sawdust was low, and neither of them was able to reach a metal uptake capacity exceeding 1 mg g^{-1}. The coconut shell and the *Moringa oleifera* seeds removed Cu(II) efficiently, accomplishing low equilibrium concentrations in solution, even for the higher initial Cu(II) concentrations (up to 150 mg L^{-1}) used in the experiments. Hence coconut shell proved to be a good sorbent for Cu(II), as evident from the high uptake of 53.9 mg g^{-1} at equilibrium concentration of 9.85 mg L^{-1} (Fig. 4-1A). The *Moringa oleifera* seeds also showed good Cu(II) removal efficiency, reaching a metal uptake of approximately 2 mg g^{-1} for a Cu(II) equilibrium concentration of 7 mg L^{-1} (Figure 4-1D).

4.4.3   Effect of particle size

Table 4-2 indicates that the equilibrium concentration of Cu(II) after sorption onto coconut shell decreases with increasing particle size. This can be attributed to the increase of the effective specific surface area with decreasing biosorbent particle size (Sengil et al., 2009). The Cu(II) removal percentage increased from 86.2 to 93.6% by decreasing the biosorbent particle diameter by a factor of 3. This suggests that Cu(II) removal by coconut shell is not considerably affected by the particle size.

4.4.4   Kinetics of Cu(II) sorption by coconut shell

The Cu(II) removal increases rapidly within the first 3 hours, reaching 70%, and then became gradually steady during the last 7 hours of contact time, where a maximum removal of 85% of the Cu(II) in solution was achieved (Figure 8A). Hence, this study has shown that the equilibrium time for the sorption of Cu(II) by coconut shell is 10 hours.   The pseudo second-order equation fit the experimental data well with correlation coefficient that is close to unity ($R^2 = 0.997$), as compared to the first-order model ($R^2 = 0.819$). The deviation from the straight line in the first 30 min of sorption, observed for the pseudo first-order model, is attributed to the fast initial uptake of Cu(II) onto the biosorbents prior to the slow biosorption in the pores. Contrary to the first-order model, the pseudo-second-order model predicts the sorption behavior over the whole time of biosorption (Ho and MacKay, 1999). On the basis of the correlation coefficients, the biosorption of Cu(II) by coconut shell follows a second-order reaction pathway.

4.4.5   Mechanism of Cu(II) biosorption

Many processes such as surface complexation, chemical precipitation, physical adsorption and ion exchange can contribute to biosorption. Recently, the dominating role of the latter process was confirmed (Chojnacka, 2010). In the ion exchange mechanism, protons compete with metal cations for binding site. FTIR (Figure 4-5), SEM-EDX (Table 4-5) and PZC (Figure 4-4) analysis results confirmed the involvement of ion exchange and electrostatic interaction in the biosorption mechanisms.

The adsorption capacity is influenced strongly by the surface structures of carbon-oxygen-hydrogen functional groups and surface behaviour of carbon (Rahman and Islam, 2009; Sarioglu et al., 2009). Wood, coconut shell and husk are made up of three dominant components, namely, cellulose, lignocellulose and lignin with varieties of functional groups as seen in Figure 4-5. *Moringa oleifera* seeds on the other hand contain proteins with a net positive charge at a pH below 8.5 (Kumar et al., 2006). These positively charged proteins are active site for binding of metal ions.

From the SEM-EDX analysis results (Table 4-5), the release of magnesium and potassium, initially fixed onto the coconut shell, was followed simultaneously by sorption of Cu(II). Also magnesium and potassium present on the *Moringa oleifera* seeds were replaced after sorption of Cu(II). This release depends on the initial Cu(II) concentration of the solution, which implies a fixation mechanism by ion exchange. The exchange between Mg(II), K(I) and Cu(II) on *Moringa oleifera* seeds and between K(I) and Cu(II) on coconut shell indicated the involvement of ion exchange mechanisms in the biosorption process for the removal of copper by both biosorbents. The FTIR spectra analysis revealed the presence of a large number of functional groups that are capable of Cu(II) removal by means of ion exchange. The observed spectra shifts, disappearance and increase in intensity of some peaks after Cu(II) sorption indicates an interaction between Cu(II) and the biosorbents studied: -OH, - C=O and carboxylic groups were involved in Cu(II) removal by coconut shell, coconut husk and sawdust whereas -C=O, amino groups and -NH$_2$ were involved in sorption by *Moringa oleifera* seeds. This suggests that these biomaterials remove Cu(II) through an ion exchange mechanism.

From Figure 4-4, the different values of PZC of the studied biosorbents indicate that these materials have surface functional groups with varying acid and basic characteristics. PZC is useful in explaining the sorption mechanism in that at solution pHs higher than pH$_{PZC}$, the sorbent surface is negatively charged and can interact with positive metal species while at pHs lower than pH$_{PZC}$, the sorbent surface is positively charged and can interact with negative species. Hence, with the exception of sawdust (pH$_{PZC}$ = 7.2), the other three biosorbents have a pH$_{PZC}$ < pH$_{solution}$ and are therefore able to attract Cu(II) by the net negative surface charges. Electrostatic attractions may be involved in Cu(II) sorption.

It should be noted that precipitation may play a role in Cu(II) removal by coconut shell. Due to the low buffering capacity of coconut shell (Figure 4-4A); the experiments with coconut shell material were most susceptible to a pH increase and subsequent Cu(OH)$_2$ formation. However, it can be said that the precipitation mechanism is catalyzed by the sorbent, since this behaviour was only observed with coconut shell. Copper forms insoluble precipitates with a large number of anions; however, the only anions present during the experiments were Cl$^-$ and OH$^-$. CuCl$_2$ is well soluble, but the solubility product (K$_s$) of Cu(OH)$_2$ is only 1.6×10^{-19}. Therefore copper hydroxides can potentially be formed, especially at the high dissolved Cu(II) concentrations at the start of the experiment or if the pH exceeds 7. It is likely that Cu(OH)$_2$ are initially formed, but as the Cu(II) concentration drops below the maximum solubility of Cu(OH)$_2$ due to sorption, the Cu(OH)$_2$ can be expected to dissociate again. The pH in the bulk solution was monitored and stayed between 7.0 and 7.5. Moreover, SEM results did not indicate any precipitation on the surface of the coconut shell and *Moringa oleifera* seed (Fig. 4-6). However, the pH inside the pores might have been different. Therefore, the contribution of Cu(OH)$_2$ formation to the observed Cu(II) removal (Figure 4-1A) cannot be excluded, especially in the sorbent pores.

4.4.6   Application potential

From this study, coconut shell has the highest sorption capacity (q$_e$ = 53.9 mg g^{-1} at C$_e$ = 9.85 mg L^{-1}) as compared to the other studied biosorbents (Figure 4-1). It is therefore a potential industrial biosorbent. The concept of biosorption includes concentrating a sorbate in biomass, in this study, Cu(II) in coconut shell, coconut husk sawdust and *Moringa oleifera* seeds. The biological material laden with sorbate is then regenerated and reused whereas the sorbate is recovered from the eluant. In order to develop a technology using the biosorption processes described in this study, additional work on the regeneration step is required. Furthermore, laboratory column and subsequent pilot scale studies are necessary to evaluate the application of the coconut shell at industrial scale. In addition, the removal of copper from real gold mining wastewater has to be evaluated during the pilot study.

Although kinetic studies showed good fitting of the kinetic data for Cu(II) sorption onto coconut shell by the pseudo second-order kinetics model, a comprehensive kinetic modelling involving other models is required to determine which model describes the sorption system best. Furthermore, for a fixed-bed application, the role of intraparticle diffusion and mass transfer has to be evaluated.

Finally, for the application of the coconut shell and *Moringa oleifera* seeds as a biosorbent at the industrial level, comparison of the process with competing technologies such as conventional ion exchange or reverse osmosis with particular impact on the costs of biosorbents formulation into ready-to-use product are necessary.

4.5   Conclusions

This study shows that coconut shell can be an important low-cost biosorbent for Cu(II) removal. A Cu(II) uptake of 53.9 mg g^{-1} was achieved at an equilibrium concentration of 9.85 mg L^{-1}. The sorption isotherms of coconut husk, *Moringa oleifera* seeds and sawdust correlate with the Freundlich model. It reasonably described the sorption of Cu(II) by the coconut shell. The other sorption models used in this study could not predict the copper removal by coconut shell, suggesting that other processes, such as copper hydroxide precipitation, may contribute to the copper removal. The FTIR analysis results (Fig. 4-5) indicated that hydroxyl and carboxylic functional groups were involved in Cu(II) removal by coconut shell, coconut husk and sawdust through the ion exchange mechanism and that mainly amino functional groups were involved in Cu(II) removal by *Moringa oleifera* seeds. The disappearance of magnesium and potassium originally present on the coconut shell and *Moringa oleifera* seeds, indicates that ion exchange contributes to the Cu(II) uptake. The PZC results showed that electrostatic forces played a role in the sorption of Cu(II) in this study.

4.6   Acknowledgements

The authors acknowledge funding from the Netherlands Government under the Netherlands Fellowship Programme (NFP), NUFFIC award (2009-2013), Project Number: 32022513. We also acknowledge funding from the Staff Development and Postgraduate Scholarship Scheme (Kumasi Polytechnic, Ghana), the UNESCO-IHE Partner Research Fund, UPaRF III research project PRBRAMD (No. 101014) and King Abdullah University of Science and Technology (Saudi Arabia) (Award No. KUK-C1-017-12). The authors thank Johan Groen of Delft Solid Solutions (Delft, The Netherlands) for his contribution to the physical characterisation of the materials. Finally, we wish to thank Dr. Natalia Chubar of Utrecht University (The Netherlands) for her support with the FTIR analyses.

4.7 References

Abdelwahab, O., 2007. Kinetics and isotherm studies of Cu(II) removal from wastewater using various adsorbents. Egyptian Journal of Aquatic Research 33(1), 125-143.

Acemioğlu, B., Alma, M.H., 2004. Sorption of copper (II) ions by pine sawdust. European Journal of Wood and Wood Products 62, 268-272.

Acheampong, M.A., Meulepas, R.J., Lens, P.N., 2010. Removal of heavy metals and cyanide from gold mine wastewater. Journal of Chemical Technology & Biotechnology 85, 590-613.

American Public Health Association, 1995. *Standard Methods for the Examination of Water and Wastewater*, 19th Ed.; pp 1108, APHA, AWWWA, WEF, Washington, D.C., USA, (1995).

Amuda, O.S., Giwa, A.A., Bello, I.A., 2007. Removal of heavy metal from industrial wastewater using modified activated coconut shell carbon. Biochemical Engineering Journal 36, 174-181.

Arief, V.O., Trilestari, K., Sunarso, J., Indraswati, N., Ismadji, S., 2008. Recent Progress on Biosorption of Heavy Metals from Liquids Using Low Cost Biosorbents: Characterization, Biosorption Parameters and Mechanism Studies. CLEAN – Soil, Air, Water 36, 937-962.

Ayawei, N., Horsfall, Jr., M., Spiff, A.I., 2005. *Rhizophora mangle* waste as adsorbent for metal ions removal from aqueous solution. Eur. J. Sci. Res. 9, 6-11.

Ayoob, S., Gupta, A.K., Bhakat, P.B., 2007. Analysis of breakthrough developments and modeling of fixed bed adsorption system for As(V) removal from water by modified calcined bauxite (MCB). Separation and Purification Technology 52, 430-438.

Bozic, D., Stankovic, V., Gorgievski, M., Bogdanovic, G., Kovacevic, R., 2009. Adsorption of heavy metal ions by sawdust of deciduous trees. Journal of Hazardous Materials 171, 684-692.

Chojnacka, K., 2010. Biosorption and bioaccumulation - the prospects for practical applications. Environment International 36, 299-307.

De Faria, L.A., Prestat, M., Koenig, J.F., Chartier, P., Trasatti, S., 1998. Surface properties of Ni+Co mixed oxides: a study by X-rays, XPS, BET and PZC. Electrochimica Acta 44, 1481-1489.

Dundar, M., Nuhoglu, C., Nuhoglu, Y., 2008. Biosorption of Cu(II) ions onto the litter of natural trembling poplar forest. Journal of Hazardous Materials 151, 86-95.

Fiol, N., Villaescusa, I., 2009. Determination of sorbent point zero charge: usefulness in sorption studies. Environmental Chemistry Letters 7, 79-84.

Gadd, G.M., 2009. Biosorption: critical review of scientific rationale, environmental importance and significance for pollution treatment. Journal of Chemical Technology & Biotechnology 84, 13-28.

Ghana EPA, 2010. Environmental performance rating and disclosure: report on the performance of mining and manufacturing companies. Environmental Protection Agency, Accra, Ghana.

Gregg, S.J., Sing, K.S.W., 1982. Adsorption, *Surface Area and Porosity*, 2nd ed., Academic Press, London.

Gulnaz, O., Saygideger, S., Kusvuran, E., 2005. Study of Cu(II) biosorption by dried activated sludge: effect of physico-chemical environment and kinetics study. Journal of Hazardous Materials 120, 193-200.

Habib, A., Islam, N., Islam, A., Alam, A.M.S., 2007. Removal of Copper from Aqueous Solution Using Orange Peel, Sawdust and Bagasse. Pak. J. Anal. Environ. Chem. 8(1 & 2), 21- 25.

Ho, Y.S., John Wase, D.A., Forster, C.F., 1995. Batch nickel removal from aqueous solution by sphagnum moss peat. Water Research 29, 1327-1332.

Ho, Y.S., McKay, G., 1999. Pseudo-second order model for sorption processes. Process Biochemistry 34, 451-465.

Horsfall, Jr. M., Spiff, A.I., Abia, A.A., 2004. Studies on the influence of mercaptoacetic acid (MAA) modification of cassava (Manihot sculenta Cranz) waste biomass on the adsorption of Cu^{2+} and Cd^{2+} from aqueous solution. Bulletin of the Korean Chemical Society 25, 969-976.

Igwe, J.C., Abia, A.A., 2007. Equilibrium sorption isotherm studies of Cd(II), Pb(II) and Zn(II) ions detoxification from waste water using unmodified and EDTA-modified maize husk. Electronic Journal of Biotechnology 10, 536-548.

Kalavathy, M.H., Karthikeyan, T., Rajgopal, S., Miranda, L.R., 2005. Kinetic and isotherm studies of Cu(II) adsorption onto H3PO4-activated rubber wood sawdust. Journal of Colloid and Interface Science 292, 354-362.

Kumar, Y.P., King, P., Prasad, V.S.R.K., 2006. Zinc biosorption on Tectona grandis L.f. leaves biomass: Equilibrium and kinetic studies. Chemical Engineering Journal 124, 63-70.

Kumaressan, M., Riyazuddin, P., 2001. Overview of speciation chemistry of arsenic. Current Science 80(7), 837-846.

Kumari, P., Sharma, P., Srivastava, S., Srivastava, M.M., 2006. Biosorption studies on shelled Moringa oleifera Lamarck seed powder: Removal and recovery of arsenic from aqueous system. International Journal of Mineral Processing 78, 131-139.

Langmuir I., 1916. The adsorption of gases on plain surface of glass, mica and platinum. J. Am. Chem. Soc. 40, 1361-1368.

Mahamadi, C., Nharingo, T., 2010. Utilization of water hyacinth weed (*Eichhornia crassipes*) for the removal of Pb(II), Cd(II) and Zn(II) from aquatic environments: an adsorption isotherm study. Environmental Technology 31, 1221 - 1228.

Pino, G.H., de Mesquita, L.M.S., Torem, M.L., Pinto, G.A.S., 2006. Biosorption of Heavy Metals by Powder of Green Coconut Shell. Separation Science and Technology 41, 3141-3153.

Rahman, M.S., Islam, M.R., 2009. Effects of pH on isotherms modeling for Cu(II) ions adsorption using maple wood sawdust. Chemical Engineering Journal 149, 273-280.

Sajidu, S.M.I., Persson, I., Masamba, W.R.L., Henry, E.M.T., 2008. Mechanisms for biosorption of chromium(III), copper(II) and mercury(II) using water extracts of Moringa oleifera seed powder. African Journal of Biotechnology 7, 800-804.

Sarioglu, M., Güler, U.A., Beyazit, N., 2009. Removal of copper from aqueous solutions using biosolids. Desalination 239, 167-174.

Sengil, I.A., Özacar, M., Türkmenler, H., 2009. Kinetic and isotherm studies of Cu(II) biosorption onto valonia tannin resin. Journal of Hazardous Materials 162, 1046-1052.

Singh, T.S., Pant, K.K., 2004. Equilibrium, kinetics and thermodynamic studies for adsorption of As(III) on activated alumina. Separation and Purification Technology 36, 139-147.

Tsai, W.T., Yang, J.M., Lai, C.W., Cheng, Y.H., Lin, C.C., Yeh, C.W., 2006. Characterization and adsorption properties of eggshells and eggshell membrane. Bioresource Technology 97, 488-493.

Ulmanu, M., Marañón, E., Fernández, Y., Castrillón, L., Anger, I., Dumitriu, D., 2003. Removal of copper and cadmium ions from diluted aqueous solutions by low cost and waste material adsorbents. Water, Air, and Soil Pollution 142, 357-373.

Vijayaraghavan, K., Padmesh, T.V.N., Palanivelu, K., Velan, M., 2006. Biosorption of nickel(II) ions onto Sargassum wightii: Application of two-parameter and three-parameter isotherm models. Journal of Hazardous Materials 133, 304-308.

Wang, J., Chen, C., 2009. Biosorbents for heavy metals removal and their future. Biotechnology Advances 27, 195-226.

Chapter 5

5 Cyclic sorption and desorption of Cu(II) onto coconut shell and iron oxide coated sand

This chapter has been presented and published as:

Dapcic, A.D., **Acheampong, M.A.,** Lens, Piet N.L., 2012. Removal of Cu(II) from Aqueous Solution Using Coconut Shell and Iron Oxide Sand: Sorption and Desorption Cycle Studies. In: *Proceedings of the 2012 University of South Florida Undergraduate Research Colloquium,* Florida, USA (18 April, 2012).

Acheampong, M.A., Dapcic, A.D., Yeh, D., Lens, P.N.L., 2012. Cyclic sorption and desorption of Cu(II) onto coconut shell and iron oxide coated sand. Separation Science and Technology (DOI:10.1080/01496395.2013.809362).

Abstract

Sorption is a viable treatment technology for copper-rich gold mine tailings wastewater. For continuous application, the sorbent should be regenerated with an appropriate desorbent, and reused. In this study, the sequential sorption/desorption characteristics of Cu(II) on coconut shell (CS) and iron oxide coated sand (IOCS) were determined. In batch assays, CS was found to have a Cu(II) uptake capacity of 0.46 mg g^{-1} and yielded a 93% removal efficiency, while the IOCS had a Cu(II) uptake capacity and removal efficiency of 0.49 mg g^{-1} and 98%, respectively. Desorption experiments indicated that HCl (0.05 M) was an efficient desorbent for the recovery of Cu(II) from CS, with an average desorption efficiency of 96% (sustained for eight sorption and desorption cycles). HCl (0.05 M) did not diminish the CS's ability to sorb copper, even after eight sorption/desorption cycles, but completely deteriorated the iron oxide structure of the IOCS within six cycles. This study showed that CS and IOCS are both good sorbents for Cu(II); but cyclical sorption/desorption using 0.05 M HCl is only feasible with CS.

Keywords: coconut shell; iron oxide coated sand; sorption; desorption; kinetics; copper; gold mining wastewater

5.1   Introduction

Industrialization can lead to great economic gain, yet it is frequently accompanied by negative environmental and health effects caused by byproducts of factories and exploitation of natural resources. Gold mining products comprise 90% of Ghana's mineral exports, and are essential to the Ghanaian economy (www.anglogold.com, assessed: 11/04/2012). However, mining contributes significantly to the spillage and discharge of contaminated wastewater into rivers and other aquatic systems, threatening human life, flora and fauna (Acheampong et al., 2010). Copper-bearing mining wastes and acid mine drainages discharge significant quantities of dissolved copper into the environment (Rahman and Islam, 2009; Chojnacka, 2010). Other potential sources of copper emissions into the environment include industrial effluents, such as metal cleaning and plating baths, pulp and paper board mills, wood pulp production, and the fertilizer industry (Ofomajo et al., 2010).

Copper and arsenic are the most toxic heavy metals present in Ghana's gold mine tailings wastewater, with average concentrations of 5 mgL^{-1} and 10 mgL^{-1}, respectively (Acheampong et al., 2011a). Coconut shell (CS) was shown to be a good biosorbent for Cu(II) (Acheampong et al., 2011a) (Chapter 4). Iron oxide coated-sand (IOCS), typically used for arsenic sorption (Gupta et al., 2005; Hsu et al., 2008), could be a potential low-cost sorbent for divalent metals, such as copper, as well.

One of the main attributes of biosorption is the potential of sorbent regeneration and metal recovery (Gupta and Rastogi, 2008; Liu et al., 2010). However, most published work has aimed to evaluate the binding ability of biomass and factors affecting the binding process. In contrast, little attention is paid to the regeneration ability of biosorbents, which is often the deciding factor on the industrial applicability of a biosorption process. Hence, to improve process viability, better understanding of the sorption-desorption processes in sorbent regeneration is needed.

Desorption of sorbed metals from loaded sorbent enables reuse of the sorbent, and recovery and/or containment of the sorbed metals from a highly concentrated eluant stream (Gupta and Rastogi, 2008; Gadd, 2009). In some cases, desorption treatment may further improve the sorption capacities (Liu et al., 2002); although in other cases there may be a loss of sorption efficiency and even a complete disintegration of the biosorbent (Marin and Ayele, 2008). A successful desorption process requires proper selection of the eluent, which strongly depends on the type of sorbent and the sorption mechanism (Acheampong et al., 2010) (Chapter 2). Ideally, the eluent is effective, low-cost, environmentally-friendly, and non-damaging to the sorbent. A variety of eluents have been suggested as metal desorbents, including acids (Kumar et al., 2007; Gadd, 2009), alkalines (Kumar et al., 2007) and chelating agents (Mata et al., 2010) depending on the substances sorbed, process requirements and economic considerations (Akar and Tunali, 2005). Effective desorption eluents can be proton exchangers, complexing agents or contain competing ions (Javanbakht et al., 2011).

In our recent publications, the sorption properties (Acheampong et al., 2011a) and kinetics modelling of Cu(II) biosorption (Acheampong et al., 2011b) onto coconut shell and other agricultural materials were studied (Chapters 4 and 6). In this work, sorption-desorption cycles and the kinetics of the desorption process of Cu(II) using coconut shell were studied under batch conditions. These were compared to those of the well characterised iron oxide coated sand (IOCS) sorbent which is used in practice for arsenic removal (Gupta et al., 2005).

5.2   Materials and Methods

5.2.1   Sorbents
The coconut shell (CS) was obtained from Kumasi (Ghana) and was described in detail by Acheampong et al. (2011a). The material was washed with distilled water and dried at 105 ^{0}C for 24 h. Shells were ground using a Peppink hammer mill, then sieved to obtain the 0.5–0.8 mm grain size used in this study. The iron oxide-coated sand (IOCS) was obtained from the water treatment plant at Zwolle Engelse Werk (The Netherlands) as a spent product of the sand filtration process. The IOCS was used as received, with a particle size range of 1.0 - 3.0 mm. Care was taken not to destroy the coating on the sand particles before and during the experiments Table 5-1 gives the physical characteristics of both sorbents used.

Table 5-1: Physical characterisation of the sorbents used in this study

Sorbent	Total pore volume ($cm^3 g^{-1}$)	Porosity (%)	Bulk density ($g\ cm^{-3}$)	Surface area (S_{BET}) ($m^2 g^{-1}$)	Reference
CS	0.900	55	1.35	0.4	Acheampong et al. (2011a)
IOCS	0.012	51	2.38	11.4	This study

5.2.2   Metal ion and desorption solutions
All reagents used in this study were of analytical grade. A stock solution (100 mg L^{-1}) of Cu(II) was prepared by dissolving the appropriate amount of copper chloride (CuCl$_2$) in demineralised water. Sorption solutions (10 mg L^{-1} Cu(II)) were prepared by diluting the appropriate volume of the stock solution in demineralised water. Solutions of hydrochloric acid, EDTA, sodium hydroxide, acetate and calcium nitrate (all at 0.2 M) were prepared and used for the desorption experiments. Sodium hydroxide (NaOH) and nitric acid (HNO$_3$) were used for pH adjustment.

5.2.3   Sorption experiments

Studies on the copper sorption cycle were carried out by agitating 100 mL of 10 mg L^{-1} metal ion solutions of Cu(II) with 2 g (dry weight) of sorbents in a 250 mL volumetric flasks at 30 ± 1 ^{0}C. The concentration was selected to reflect the copper concentration in the actual gold mining wastewater under study, which was below 10 mg L^{-1} in all the cases analysed (Acheampong et al., 2013) (Chapter 3). At the predetermined time, samples were drawn, filtered using 0.45 μm filter paper (Schleider & Schuell, No. 595 ½), acidified to a pH 2.0 using 65% HNO$_3$ and analyzed for their metal ion concentrations using atomic absorption spectrometry (AAS).

5.2.4   Desorption experiments

Prior to desorption of the metal ions from the sorbent (after completing the sorption step), the metal-loaded sorbent was washed according to a strict protocol. First, any residual solution from the sorption step was carefully drained, ensuring no sorbent was lost. The flask was then filled with demineralised water (400 mL) and drained, again ensuring no sorbent was lost. The flasks were filled and drained a total of 3 times to ensure that only bound copper remained on the sorbents. Desorption of the sorbed metal ions from the loaded sorbents were carried out by agitating 100 mL of the desorption solution with the loaded sorbent in a 250 mL volumetric flask at 30 ± 1^{0}C. At the predetermined time, liquid samples were drawn, filtered using 0.45 μm filter paper (Schleider & Schuell, No. 595 ½), and acidified to a pH 2.0 using 65% HNO$_3$. Samples were analyzed for Cu(II) concentrations using AAS. After the desorption step, the sorbent was carefully washed to remove the desorption solution and then re-entered in the next sorption/desorption cycle.

To study the desorption ability of both CS and IOCS, five different desorption solutions were screened under batch conditions using the loaded sorbent (all at a 0.2 M concentration): hydrochloric acid (HCl), ethylenediaminetetraacetic acid (EDTA), sodium hydroxide (NaOH), sodium acetate (C$_2$H$_3$NaO$_2$) and calcium nitrate (Ca(NO$_3$)$_2$). In this desorbent screening, two cycles of twenty-four hour sorption and twenty-four hour desorption were carried out in triplicate. Samples were taken at the end of each sorption and desorption step and analysed for the remaining copper concentration in solution. Since the eluant concentration is much higher than the Cu(II) concentration in the elution solution, no pH change is expected during the desorption process. Hence, no pH adjustment is required.

Following the screening, the five most efficient combinations of sorbent and desorption solution were selected for sorption/desorption cycle studies. Each combination was tested in triplicate for a minimum of five cycles of twenty-four hour sorption and twenty-four hour desorption to test the desorbent's ability to desorb, at what rate, and to examine the structure of the sorbent material. Liquid samples were drawn at the conclusion of each sorption and desorption step, and analysed for the remaining Cu(II) concentration in solution.

The effect of the desorption solution concentration was determined with the most efficient combination of sorbent and desorption solution (CS and HCl). The HCl concentration was decreased from 0.2 M to 0.05 M. At the end of each sorption and desorption step, liquid samples were taken and analysed with AAS for the dissolved Cu(II) concentration in solution.

To determine the rate at which sorption and desorption is achieved, a batch test was sampled during a sorption and desorption cycle, each lasting twenty-four hours. One cycle of sorption and desorption was completed in duplicate. Samples were then analysed to determine the remaining dissolve Cu(II) concentration.

5.2.5   Analytical techniques
The Cu(II) concentration in the filtrate was measured on an AAS (Perkin Elmer, model AAnalist200), equipped with an air-acetylene flame. The pH was measured with a Metrohm Herisau, Prazisions-pH-Meter E 510. The pH meter was calibrated using buffer solutions with pH values of 4.0, 7.0 and 10.0.

5.2.6   Calculations
The specific metal uptake (q), the percentage removal (Removal (%)), and the desorption efficiency (η_d) were calculated according to the following equations (Kumar et al., 2007):

$$q_e = \frac{C_0 - C_e}{m} \times V \qquad\qquad (5\text{-}1)$$

$$removal(\%) = \frac{C_0 - C_e}{C_0} \times 100 \qquad\qquad (5\text{-}2)$$

$$\eta_d(\%) = \frac{C_{fd}}{C_0 - C_{fs}} \times 100 \qquad\qquad (5\text{-}3)$$

where C_0 (mg L^{-1}) is the initial concentration of Cu (II) in solution, C_e (mg L^{-1}) the equilibrium Cu(II) concentration, C_{fs} (mg L^{-1}) the final Cu(II) concentration in solution after sorption, C_{fd} (mg L^{-1}) the final Cu(II) concentration in solution after desorption, q_e (mg g^{-1}) the equilibrium uptake capacity and m (g L^{-1}) the dry weight of the adsorbent.

5.3   Results

5.3.1   Equilibrium uptake of Cu(II) by CS and IOCS
Table 5-2 shows the uptake and removal (%) of Cu(II) by CS and IOCS after 24 and 72 h contact time. Both sorbents show a good affinity for Cu(II). The results show, however, that the Cu(II) removal from solution does not further increase significantly after 24 h contact time.

Table 5-2: Uptake of Cu(II) by coconut shell and IOCS. Experimental conditions: initial Cu(II) concentration = 10 mg L^{-1}, sorbent concentration = 20 g L^{-1}, particle size = 0.5 - 0.8 mm, contact time = 24 h, mixing rate = 100 rpm, temperature = 30 ± 1 ^{0}C, pH =7.0 ± 0.2

Sorbent	Uptake (mg g^{-1})		Removal (%)	
	24 h	72 h	24 h	72h
Coconut shell	0.46 (±0.0002)	0.46 (±0.0008)	92.7 (±0.0018)	92.6 (±0.0015)
IOCS	0.49 (±0.0004)	0.49 (±0.0001)	97.1 (±0.0007)	97.7 (±0.0001)

5.3.2   Screening of desorption solutions

Five potential desorption solutions were studied for their ability to desorb Cu(II) from loaded CS and IOCS (Figure 5-1). The removal efficiency of Cu(II) from solution in the second sorption step was only slightly affected after the exposure to a desorption solution. While HCl and EDTA performed well in desorbing Cu(II) from the loaded sorbents, acetate and sodium hydroxide performed poorly (Figure 5-1). Figure 5-1 further shows that HCl is the best desorbing agent for Cu(II) loaded onto the sorbents studied. At the end of the second sorption and desorption cycle, the amount of Cu(II) desorbed from loaded CS increased from 5.9 mg L^{-1} to 9.7 mg L^{-1}, while that of the IOCS increased from 5.5 mg L^{-1} to 8.5 mg L^{-1}.

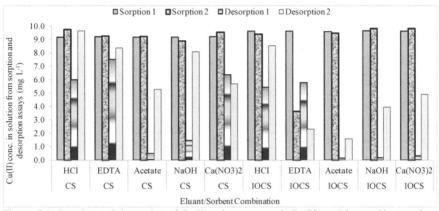

Figure 5-1: Sorption and desorption of Cu(II) using coconut shell (CS) and iron oxide coated sand (IOCS) as sorbents with HCl, EDTA, acetate, NaOH and Ca(NO$_3$)$_2$ as desorption solutions. Experimental conditions: initial Cu(II) concentration = 10 mg L^{-1}, sorbent concentration = 20 g L^{-1}, contact time = 24 h, mixing speed = 100 rpm, temperature = 30 ± 1 ^{0}C.

The HCl, EDTA, Ca(NO$_3$)$_2$ and C$_2$H$_3$NaO$_2$ did not have any noticeable physical effect on the CS, as the solution remained clear and the structure of the sorbent remained unaltered. However, NaOH dissolved part of the CS, turning both the sorbent and the solution black in colour (Figure 5-2C).

The CS particles were more stable upon exposure to 0.2 M HCl compared to the IOCS particles. The IOCS showed a considerable level of disintegration during the second sorption and desorption cycle with 0.2 M HCl (Figure 5-2D). The IOCS lost its integrity by HCl exposure, yet the stripped iron oxide coating did not fully dissolve into solution. The coating particles remained solid and discrete alongside the sand grains on which they were originally coated (Figure 5-2D). EDTA, however, quickly dissolved the iron oxide coating completely during the desorption process. Therefore, no visible iron oxide coating remained on the sand grains (Figure 5-2E). Sodium acetate, sodium hydroxide and calcium nitrate were not efficient in desorbing Cu(II) ions from the IOCS. The C$_2$H$_3$NaO$_2$ and Ca(NO$_3$)$_2$ did not have any significant physical effect on the IOCS, as the solution remained largely clear and the sorbent retained its structural integrity. Since dramatic sorbent degradation occurred with EDTA/IOCS and HCl/IOCS, these were not further investigated. HCl in combination with CS showed good prospect for sorption/desorption applications for Cu(II) removal from aqueous solutions and was hence further investigated to optimise the HCl desorption process.

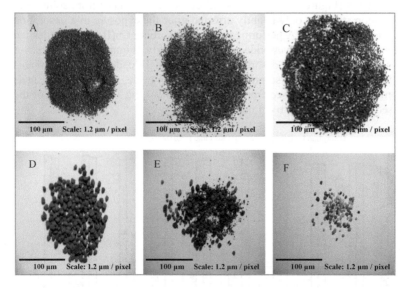

Figure 5-2: Effect of desorbent on particle size of the coconut shell (A, B, C) and IOCS (D, E, F); Prior to sorption and desorption application (A, D), after cyclical sorption and desorption with 0.05 M HCl (B, E), after cyclical sorption and desorption with NaOH (C) and EDTA (F).

5.3.3   Effect of HCl concentration on Cu(II) desorption

Figure 5-3 shows the sorption of Cu(II) onto CS as a function of residual Cu(II) concentration and time. The results of desorption of Cu(II) loaded onto CS at different eluant concentrations is presented in Figure 5-3. Figure 5-3 shows that sorption and desorption occurred rapidly and that Cu(II) was sorbed and desorbed within the first 360 min. In fact, the Cu(II) concentration in solution reduced from 10 mg L^{-1} to 1.42 mg L^{-1} within the first 60 min of sorption (Figure 5-3). The amount of Cu(II) desorbed after 30 min operation was already 8.1 mg L^{-1} with all three HCl concentrations investigated (Figure 5-3). The average desorption efficiency amounted to 96% (±0.01) for all three cases. Furthermore, Fig. 3 indicates that there is no significant difference in the amount Cu(II) desorbed (8.7 mg L^{-1}) by 0.05 M, 0.1 M and 0.2 M HCl solutions after 360 min. Consequently, 0.05 M HCl was selected for the subsequent cyclical sorption and desorption studies.

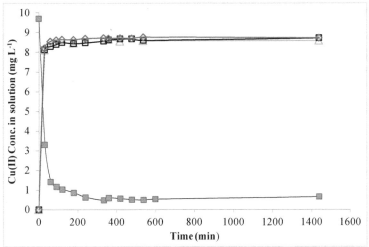

Figure 5-3: Sorption (■) and desorption kinetics of Cu(II) from coconut shell using 0.2 M HCl(◊), 0.1 M HCL (□) and 0.05 M HCl (Δ). Experimental conditions: initial Cu(II) concentration = 10 mg L^{-1}, sorbent concentration = 20 g L^{-1}, contact time = 24 h, mixing rate = 100 rpm, temperature = 30 ± 1 ^{0}C, pH = 7.0 ± 0.2 (for sorption).

5.3.4   Cyclical Cu(II) sorption and desorption studies

Figure 5-4 shows the sorption and desorption of Cu(II) over an 8 cycle period with 0.05 M HCl and CS, as well as the sorption and desorption of Cu(II) over a 6 cycle period with 0.05 M HCl and IOCS. Figure 5-5 shows the desorption efficiency as a function of desorption cycle.

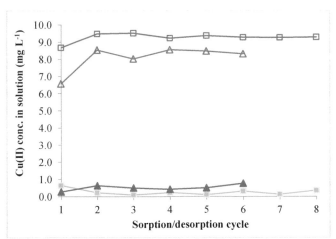

Figure 5-4: Sorption and desorption cycle of Cu(II) biosorption onto coconut shell (■ and □) and IOCS (▲ and Δ). Experimental conditions: initial Cu(II) concentration = 10 mg L^{-1}, sorbent concentration = 20 g L^{-1}, particle size = 0.5 - 0.8 mm, mixing rate = 100 rpm, temperature = 30 ± 1 ^{0}C, pH = 7.0 ± 0.2.

An average Cu (II) uptake capacity of 0.48 ± 0.01 mg g^{-1} and desorption efficiency of 95% (±0.03) was obtained with 0.05 M HCl during the 8 cycles of sorption and desorption (Figure 5-5). The first two sorption and desorption cycles agreed with earlier experimentation; specifically, there was an increased sorption efficiency after the first cycle of sorption and desorption (Figure 5-4). For the remaining cycles, sorption and desorption results were very consistent. The desorption performance of the 0.05 M HCl on the CS was better than that recorded for the IOCS under the same conditions (Figure 5-5). Though the first sorption step with IOCS produced a lower equilibrium Cu(II) concentration than recorded for the CS, the opposite occurred in the subsequent sorption steps (Figure 5-4). The 8 cycles of sorption and desorption did not appear to affect the physical structure of the CS (Figures 5-2 E and 5-2 F).

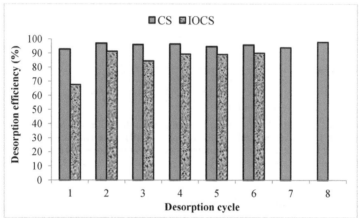

Figure 5-5: Desorption cycle of Cu(II) sorption onto coconut shell and IOCS. Experimental conditions: concentration of HCl = 0.05 M, sorbent concentration = 20 g L^{-1}, particle size = 0.5 - 0.8 mm, mixing rate = 100 rpm, temperature = 30 ± 1 ^{0}C

5.4   Discussion

5.4.1   Effect of desorption cycles on biosorbent capacity

This study showed that CS and IOCS are both well suited sorbents for Cu(II) removal from solution. IOCS exhibited a 6.5% greater equilibrium copper uptake than CS (Table 5-2), demonstrating that it is better for copper sorption in a single stage operation. Cu(II) sorption onto CS attained equilibrium in 24 h, as evident from the percentage removal over 24 and 72 h (Table 5-2). The 0.6 percentage increase in removal efficiency by IOCS is inconsequential regarding the additional time (36 h) used; therefore, equilibrium can be assumed for both sorbents within 24 h. Excluding IOCS in conjunction with EDTA, sorption efficiencies increased or remained the same from the first to second sorption cycle (Figure 5-1).

The lower equilibrium Cu(II) concentration recorded in the first sorption step with IOCS as opposed to the subsequent sorption steps (Figure 5-4) suggest an irreversible bonding of some of the Cu(II) ions loaded on to the IOCS. In contrast, the sorbed copper in the CS was potentially mobile, as judged by its ease of desorption upon treatment with 0·05 M HCl, suggesting a weak bonding between Cu(II) ions and the CS. HCl produced efficient desorption of Cu(II) from the CS in both the first and second desorption steps (Fig. 5-4), with a significant increase in efficiency in the latter process. The second sorption yielded the highest sorption efficiency of all experiments, suggesting a modification of the chemical structure of the sorbent. If the

Cu(II) ions were strongly bonded to CS, a decrease in sorption efficiency would have been apparent after the first cycle of sorption and desorption. This was supported by Acheampong et al. (2011a), who suggested that the forces of attraction are Van der Waal and not chemical bonding.

5.4.2   Effect of desorption cycles on biosorbent structure

There was no visible damage to the structure of the CS which retained its macroscopic appearance and performance in repeated metal uptake/elution cycles. Holan et al. (1993) reported similar results for desorption of cadmium loaded onto marine algae biomass by 0.1 - 0.5 M HCl. The sorption and desorption performance of dried *Ecklonia maxima* was studied by Stirk and van Staden (2000). They reported that no deterioration in its metal binding capacity was recorded after four cycles of sorption/desorption using 0.1 M HCl. According to Sivaprakash et al. (2010), copper loaded biomass was eluted using 0.1 M HCl and no damage to the biosorbent was caused after 7 cycles of sorption/desorption.

The main drawback for the IOCS was the visible structural damage caused by the desorbing HCl. This may be due to the fact that the iron oxide was loosely deposited to the surface of the sand or the dissolution of the bonds yielding its physical strength, a property which does not allow IOCS to be used in cyclical mode using strong desorption agents. Aldor et al. (1995) reported a similar structural damage to Kelpak (a commercial seaweed concentrate used as a growth stimulant in agriculture) waste when 0.1 M HCl was used to desorb Cu(II) ions. Desorption of Cu(II) ions loaded onto a new spherical cellulose adsorbent with 2.4 M HCl was studied by Liu et al. (2002). They achieved 100% desorption efficiency and reported only a 7.2% loss in adsorption capacity after 30 sorption/desorption cycles. It is important to note that similar results can be achieved with a lower concentration of HCl (0.05 M), as demonstrated in this study. In fact, desorption of over 99% copper bound onto sodium hydroxide treated rubber (*Hevea brasiliensis*) leaves powder was achieved using 0.05 M HCl, 0.01 M HNO_3 and 0.01 M EDTA (Ngah, 2008).

5.4.3   Effect of desorption solutions

This study indicated that HCl is the most effective desorbing eluent for Cu(II) ions loaded onto CS, which is in agreement with Chu and Hashim (2001), who found that HCl is an effective eluant for Cu(II) ions loaded onto marine microalgae. The other studied eluents either gave poor Cu(II) recovery, resulted in a colouration in the solution due to leaching of unspecified compounds (CS) or caused structural damage to the sorbent (IOCS). This adversely affected subsequent sorption.

Mineral acids such as HNO_3 and HCl are proton-exchange agents, and have been reported as having high dislodge capacity for metal ions from biomass (Javanbakht et al., 2011). Aldor et al. (1995) working with cross-linked *Sargassum fluitans* found that the three acids HCl, HNO_3 and H_2SO_4, the chelator Na_2EDTA and $CaCl_2$ salt were the most effective desorption eluents for Cd(II) ions. They indicated however that the high cost of the Na_2EDTA and $CaCl_2$ salt and the unspecified damage to the biosorbent material by HNO_3 and H_2SO_4 made these eluents rather unsuitable as desorbing agents. They thus selected HCl as the most potent desorbing agent (Aldor et al., 1995).

EDTA enabled the most efficient first desorption of Cu(II) ions from CS (Figure 5-1). Its high performance is likely attributed to the chelating properties of EDTA acting alongside the sorption sites on the interior and exterior of the CS. As a chelating agent, EDTA has two or more donor atoms that can simultaneously coordinate to a metal ion (Ngah, 2008). However, a loss of desorption efficiency during the second cycle (Figure 5-1) may be due to a stronger Cu(II) bonding as a result of the chemical modification of the structure of the CS by the EDTA; indicating that some of the Cu(II) ions were irreversibly bound onto the sorbent (Covelo et al., 2007). Gupta et al. (2009) found a 100% desorption of Cu(II) and Ni(II) ions from Irish peat moss using 2 mM EDTA, while Deng et al. (2007) reported that 0.01 M Na_2EDTA was an efficient desorbent for the recovery of Pb(II) from green algae biomass. According to Debnath et al. (2011), 0.01 M EDTA and 0.1 M HCl were effective in desorbing Cu(II) from nanoparticle agglomerates of hydrous titanium (VI) oxide.

Acetate did not exhibit an effective first desorption, but showed a significant increase in the amount of Cu(II) desorbed in the second desorption step (Figure 5-1). A change in the chemical structure of the CS likely occurred, allowing for increased desorption capacity. NaOH showed a similar performance as the acetate, although it was observed that NaOH dissolved part of the CS (Figure 5-2C).

$Ca(NO_3)_2$ exhibited a reduction in desorption ability between the first and second desorption cycle, although a high sorption efficiency was recorded (Figure 5-1). A similar result was published by Abat et al. (2012), who employed $Ca(NO_3)_2$ to desorb Cu(II) and Zn(II) ions from tropical peat soil. The loss of sorption performance during multiple uses may be due to a number of reasons. It may be caused by changes in the chemistry and in the structure of the biosorbents, as well as by changes in the mass transport conditions within the system (Volesky et al. 2003). Khan et al. (2007) reported a low desorption efficiency of 2.5 to 6% for Cu(II) from pasture soils in 24 h. This finding is similar to the results published by Arias et al. (2006), who reported that Cu(II) desorbed from loaded soils in no case exceeded 11% of the previously adsorbed Cu(II). Marin and Ayele (2008), on the other hand, observed a 23% decrease in sorption capacity of sawdust for Cu(II) ions between cycles 1 and 2, with protons as desorbing agent. They explained that the low desorption efficiency recorded indicates that the Cu(II) ions were irreversibly bound to the sorbents.

5.5   Conclusion

This study showed that, in a batch reactor, CS can be used in sorption and desorption cycles to treat water containing dissolved Cu(II) ions up to 10 mg L^{-1}. HCl, even after eight cycles, does not affect CS's ability to sorb Cu(II) ions nor does it deteriorate its particle structure. Therefore, the sorbent does not need to be replaced every single sorption step and the application of such a sorption/desorption cycle decreases process costs. In addition, rapid sorption and desorption times (60 minutes) add to its applicability to treat larger volumes of wastewater treatment. Though IOCS exhibited a high capacity to sorb Cu(II) ions, it cannot be used for cyclical sorption/desorption applications due to the structural damage caused by the desorbents investigated.

5.6   Acknowledgements

The authors acknowledge funding from the Dutch Government under the Netherlands Fellowship Programme (NUFFIC award 32022513), the Staff Development and Postgraduate Scholarship Scheme (Kumasi Polytechnic, Ghana), the UNESCO-IHE Partner Research Fund (2009 category III small project award, PRBAMD) and the International Research Experience for Students Program (Award Number 0854306) of the U.S. National Science Foundation. We wish to thank Mr. Fred Kruis and the UNESCO-IHE laboratory staff and Dr. Maya Trotz of the University of South Florida (USA) for their support

5.7   References

Abat, M., McLaughlin, M.J., Kirby, J.K., Stacey, S.P., 2012. Adsorption and desorption of copper and zinc in tropical peat soils of Sarawak, Malaysia. Geoderma 175–176, 58-63.

Acheampong, M.A., Paksirajan, K., Lens, P.L., 2013. Assessment of the effluent quality from gold mining industry in Ghana. Environmental Science and Pollution Research 20, 3799-3811.

Acheampong, M.A., Pereira, J.P.C., Meulepas, R.J.W., Lens, P.N.L., 2011a. Biosorption of Cu(II) onto agricultural materials from tropical regions. Journal of Chemical Technology & Biotechnology 86, 1184-1194.

Acheampong, M.A., Pereira, J.P.C., Meulepas, R.J.W., Lens, P.N.L., 2011b. Kinetics modelling of Cu(II) biosorption on to coconut shell and Moringa oleifera seeds from tropical regions. Environmental Technology 33, 409-417.

Acheampong, M.A., Meulepas, R.J., Lens, P.N., 2010. Removal of heavy metals and cyanide from gold mine wastewater. Journal of Chemical Technology & Biotechnology 85, 590-613.

Aldor, I., Fourest, E., Volesky, B., 1995. Desorption of cadmium from algal biosorbent. The Canadian Journal of Chemical Engineering 73(4), 516-522.

Akar, T., Tunali, S., 2005. Biosorption performance of Botrytis cinerea fungal by-products for removal of Cd(II) and Cu(II) ions from aqueous solutions. Minerals Engineering 18, 1099-1109.

Arias, M., Pérez-Novo, C., López, E., Soto, B., 2006. Competitive adsorption and desorption of copper and zinc in acid soils. Geoderma 133, 151-159.

Chojnacka, K., 2010. Biosorption and bioaccumulation - the prospects for practical applications. Environment International 36, 299-307.

Chu, K.H., Hashim, M.A., 2001. Desorption of Copper from Polyvinyl Alcohol-Immobilized Seaweed Biomass. Acta Biotechnologica 21, 295-306.

Covelo, E.F., Vega, F.A., Andrade, M.L., 2007. Competitive sorption and desorption of heavy metals by individual soil components. Journal of Hazardous Materials 140, 308-315.

Debnath, S., Nandi, D., Ghosh, U.C., 2011. Adsorption–Desorption Behavior of Cadmium(II) and Copper(II) on the Surface of Nanoparticle Agglomerates of Hydrous Titanium(IV) Oxide. Journal of Chemical & Engineering Data 56, 3021-3028.

Deng, L., Su, Y., Su, H., Wang, X., Zhu, X., 2007. Sorption and desorption of lead (II) from wastewater by green algae Cladophora fascicularis. Journal of Hazardous Materials 143, 220-225.

Gadd, G.M., 2009. Biosorption: critical review of scientific rationale, environmental importance and significance for pollution treatment. Journal of Chemical Technology & Biotechnology 84, 13-28.

Gupta, V.K., Rastogi, A., 2008. Sorption and desorption studies of chromium(VI) from nonviable cyanobacterium Nostoc muscorum biomass. Journal of Hazardous Materials 154, 347-354.

Gupta, V.K., Saini, V.K., Jain, N., 2005. Adsorption of As(III) from aqueous solutions by iron oxide-coated sand. Journal of Colloid and Interface Science 288, 55-60.

Holan, Z.R., Volesky, B., Prasetyo, I., 1993. Biosorption of cadmium by biomass of marine algae. Biotechnology and Bioengineering 41, 819-825.

http://www.anglogold.com/subwebs/informationforinvestors/reports09/AnnualReport09/f/AGA_AR09. pdf, Accessed: 11/04/2012.

Hsu, J.-C., Lin, C.-J., Liao, C.-H., Chen, S.-T., 2008. Removal of As(V) and As(III) by reclaimed iron-oxide coated sands. Journal of Hazardous Materials 153, 817-826.

Javanbakht, V., Zilouei, H., Karimi, K., 2011. Lead biosorption by different morphologies of fungus

Mucor indicus. International Biodeterioration & Biodegradation 65, 294-300.

Khan, M.A.R., Bolan, N.S., MacKay, A.D., 2005. Adsorption and Desorption of Copper in Pasture Soils. Communications in Soil Science and Plant Analysis 36, 2461-2487.

Kumar, G.P., Kumar, P.A., Chakraborty, S., Ray, M., 2007. Uptake and desorption of copper ion using functionalized polymer coated silica gel in aqueous environment. Separation and Purification Technology 57, 47-56.

Liu, D., Sun, D., Li, Y., 2010. Removal of Cu(II) and Cd(II) From Aqueous Solutions by Polyaniline on Sawdust. Separation Science and Technology 46, 321-329.

Liu, M., Deng, Y., Zhan, H., Zhang, X., 2002. Adsorption and desorption of copper(II) from solutions on new spherical cellulose adsorbent. Journal of Applied Polymer Science 84, 478-485.

Mata, Y.N., Blázquez, M.L., Ballester, A., González, F., Muñoz, J.A., 2010. Studies on sorption, desorption, regeneration and reuse of sugar-beet pectin gels for heavy metal removal. Journal of Hazardous Materials 178, 243-248.

Marin, J., Ayele, J., 2003. Removal of some heavy metal cations from aqueous solutions by spruce sawdust. II. Adsorption-desorption through column experiments. Environmental Technology 24, 491-502.

Ngah, W.S.W., Hanafiah, M.A.K.M., 2008. Biosorption of copper ions from dilute aqueous solutions on base treatedrubber (Hevea brasiliensis) leaves powder: kinetics, isotherm, and biosorption mechanisms. Journal of Environmental Sciences 20, 1168-1176.

Ofomaja, A.E., Naidoo, E.B., Modise, S.J., 2010. Biosorption of copper(II) and lead(II) onto potassium hydroxide treated pine cone powder. Journal of Environmental Management 91, 1674-1685.

Rahman, M.S., Islam, M.R., 2009. Effects of pH on isotherms modeling for Cu(II) ions adsorption using maple wood sawdust. Chemical Engineering Journal 149, 273-280.

Sen Gupta, B., Curran, M., Hasan, S., Ghosh, T.K., 2009. Adsorption characteristics of Cu and Ni on Irish peat moss. Journal of Environmental Management 90, 954-960.

Sharma, S.K.; Petrusevski, B.; Schippers, J. C., Characterisation of coated sand from iron removal plants. Water Science and Technology: Water Supply 2002, 2, (2), 247-257.

Sivaprakash, B., Rajamohan, N., Mohamed Sadhik, A., 2010. Batch and column sorption of heavy metal from aqueous solution using a marine alga (sargassum tenerrimum). International Journal of ChemTech Research 2 (1), 155-162.

Stirk, W.A., van Staden, J., 2002. Desorption of Cadmium and the Reuse of Brown Seaweed Derived Products as Biosorbents. Botanica Marina 45, 9-6.

Volesky, B., Weber, J., Park, J.M., 2003. Continuous-flow metal biosorption in a regenerable Sargassum column. Water Research 37, 297-306.

Chapter 6

6 Kinetics modelling of Cu(II) biosorption onto coconut shell and *Moringa oleifera* seeds from tropical regions

This was published as:

Acheampong, M.A., Pereira, J.P.C., Meulepas, R.J.W., Lens, P.N.L., 2011. Kinetics modelling of Cu(II) biosorption on to coconut shell and Moringa oleifera seeds from tropical regions. Environmental Technology 33, 409-417.

Abstract

Adsorption kinetic studies are of great significance to evaluate the performance of a given adsorbent and to gain insight into the underlying mechanism. This work investigated the sorption kinetics of Cu(II) onto coconut shell and *Moringa oleifera* seeds using batch techniques. In order to understand the mechanisms of the biosorption process and the potential rate controlling steps, kinetic models were used to fit the experimental data. The results indicate that kinetic data were best described by the pseudo second-order model with correlation coefficient (R^2) of 0.9974 and 0.9958 for the coconut shell and *Moringa oleifera* seeds, respectively. The initial sorption rates obtained for coconut shell and *Moringa oleifera* seeds were 9.6395×10^{-3} and 8.3292×10^{-2} mg g^{-1} min^{-1}, respectively. The values of the mass transfer coefficients obtained for coconut shell ($\beta_l = 1.2106 \times 10^{-3}$ cm s^{-1}) and *Moringa oleifera* seeds ($\beta_l = 8.965 \times 10^{-4}$ cm s^{-1}) indicate that the transport of Cu(II) from the bulk liquid to the solid phase was quite fast for both materials investigated. The results indicate that intraparticle diffusion controls the rate of sorption in this study; however film diffusion can not be neglected, especially at the initial stage of sorption.

Keywords: copper; biosorption; kinetics modelling; coconut shell; Moringa oleifera

6.1   Introduction

The increasing levels of toxic metals that are discharged into the environment as industrial wastes represent a serious threat to human health, living resources and ecological systems (Svilovic et al., 2009). Gold mine wastewaters generally contain heavy metal pollutants, such as copper, arsenic, iron, zinc and lead at elevated concentrations. The presence of copper in mining wastewater contributes to water pollution (Ahmady-Asbchin et al., 2008; Acheampong et al., 2010). Although copper is an essential element, acute doses cause metabolic disorders, kidney damage, liver damage or the Wilson disease (Sengil et al., 2009; Febrianto et al., 2009). Treatment of copper containing wastewater can be achieved by chemical precipitation, ultra filtration, sorption and biosorption (Svilovic et al., 2009). Biosorption is based on the metal-binding capacity of various biological materials, such as algae, bacteria, agricultural and plant materials. It presents an alternative technology for toxic metal removal from waste streams (Volesky, 2001; Beolchini et al., 2003; Gadd, 2009). Coconut shell (Goel et al., 2004) and *Moringa oleifera* seed powder (Sajidu et al., 2008) are two of the agricultural materials that have been used for the sorption of copper from aqueous solutions.

The kinetics of adsorption at solid/liquid interfaces is of crucial importance in biological systems and for a variety of technological processes (Radzinski and Plazinski, 2007). The overall sorption process may be controlled by film diffusion, intraparticle diffusion or sorption on the surface (Febrianto et al., 2009). In order to gain a better understanding of the biosorption process, various kinetic models are used to test experimental data. The pseudo first-order (Abdelwahab, 2007; Ngah and Hanafiah, 2008), pseudo second-order (Ho and MacKay, 1999; Chojnacka, 2006; Pamukoglu and Kargi, 2006) and intraparticle diffusion (Gulnaz et al., 2005; Ncibi et al., 2008) models are used frequently. Modelling of kinetic data is fundamental for the industrial application of sorption because it enables comparison among different biosorbents under different operational conditions (Anirudhan and Radhakrishnan, 2008). Modelling offers useful information to gain insight into adsorption mechanisms and to design fixed-bed systems (Qiu et al., 2009; Farooq et al., 2010). It is important in optimizing operational conditions for pollutant removal from wastewater systems.

The metal uptake rate is influenced by factors affecting mass transfer from the bulk solution to the binding sites (Vilar et al., 2007). The concentrations of the products do not appear in the rate law because the reaction rate is studied under conditions where the reverse reactions do not contribute to the overall rate. The reaction order and rate constant must be determined experimentally (Ho and Wang, 2004). Knowledge of the nature and the theoretical description of the kinetics are crucial for practical applications, as a key to design the adsorption equipment and operational conditions that give optimum efficiency (Horsfall et al., 2006). Generally, adsorption kinetics are the basis for the determination of the performance of fixed-bed or any other flow-through system (Hameed and El-Khaiary, 2008).

The prediction of the rate-limiting step is an important factor in the design of the adsorption process (Kalavathy et al., 2005). It is governed by the adsorption mechanism. For a solid–liquid sorption process, the solute transfer is usually characterized by external mass transfer (boundary layer diffusion), or intraparticle diffusion, or both. The mechanisms of adsorption can be analysed in three steps as follows (Kalavathy et al., 2005; Radzinski and Plazinski, 2007):
1. Transport of the solute from the bulk solution through the liquid film to the adsorbent exterior surface
2. Transport of the adsorbate within the pores of the adsorbent (particle diffusion)
3. Adsorption of the adsorbate on the exterior surface of the adsorbent.

Generally, the last step is the equilibrium reaction and it is very rapid; the resistance is assumed to be negligible (Rudzinski and Plazinski, 2007). The slowest step determines the rate; the rate-controlling parameter can be distributed between intraparticle and film diffusion mechanisms (Yao et al., 2010).

In our recent work (Acheampong et al., 2011), the equilibrium sorption capacity and affinity of four agricultural materials, namely, coconut shell, coconut husk, sawdust and *Moringa oleifera* seeds for Cu(II) from gold mining wastewater was studied (Chapter 4). The study showed that the biosorbents are potential low-cost materials for Cu(II) removal. Coconut shell was found to have the highest Cu(II) sorption capacity (53.6 mg g^{-1}), followed by *Moringa oleifera* seeds (2.5 mg g^{-1}).

This study describes the biosorption kinetics of Cu(II) onto coconut shell and *Moringa oleifera* seeds by fitting the experimental data onto the pseudo first-order, pseudo second-order, intraparticle diffusion, Elovich and mass transfer models. Good understanding of the mass transfer of Cu(II) from the bulk liquid across the boundary layer to the surface of the biosorbent particles and the subsequent pore diffusion is necessary for the design of fixed-bed columns for laboratory scale and pilot plant studies, and for a full-scale application of the sorption process.

6.2   Materials and methods

6.2.1   Biosorbents
The coconut shell and *Moringa oleifera* seeds were obtained from Kumasi (Ghana) and described in detail by Acheampong et al. (2011) (Chapter 4). The materials were washed with distilled water and dried at 105 ^{0}C for 24 h. The coconut shells were grinded using a Peppink hammer mill, while a laboratory blender was used to grind the *Moringa oleifera* seeds. The biosorbents were then sieved to obtain the 0.5 – 0.8 mm fraction used in this experiment. Table 1 shows the physical characteristics of the biosorbents.

Table 6-1: Physical characteristics of the biosorbents

Biosorbent	Specific* volume (cm^{3}g^{-1})	Porosity (%)	ρ (g cm^{-3})	Surface area (S$_{BET}$) (m^{2}g^{-1})
Coconut shell	0.90	55	1.35	0.4
Moringa oleifera seeds	0.46	34	1.14	0.1

*Including the pore volume

6.2.2   Metal ion solutions
All reagents used in this study were analytical grade. A stock solution (1000 mg L^{-1}) of Cu(II) was prepared by dissolving the appropriate amount of copper chloride (CuCl$_2$) in distilled water. Solutions of lower concentrations were prepared by diluting the stock solution with double distilled water. Sodium hydroxide (NaOH) and hydrochroric acid (HCl) were used for pH adjustment.

6.2.3   Batch biosorption kinetics experiments
Studies on the rate of adsorption of the metal ion were carried out by agitating 100 ml of 10 mg L^{-1} metal ion solutions of Cu(II) with 2 g (dry weight) of biosorbent (particle size range: 0.5-0.8 mm) in volumetric flasks at 30 ± 0.2 ^{0}C. Samples were taking from the flask at predetermined time intervals (every 30 min) and analyzed for metal ion concentrations using atomic absorption spectrometry (AAS).

6.2.4   Analytical techniques
The concentration of Cu(II) in the filtrate was measured on an atomic absorption spectrometer (Perkin Elmer, model AAnalist200), equipped with an air-acetylene flame. The pH was measured with a SenTix21 pH electrode (WTW model pH323). The instrument was calibrated using buffer solutions with pH values of 4.0, 7.0 and 10.0.

6.3   Sorption kinetics models and calculations

6.3.1   Sorption kinetics models
The pseudo first- and second-order, intraparticle diffusion, Elovich and mass transfer models were used.

6.3.1.1   *Pseudo first-order kinetic model*
The first-order rate expression of Lagergren (Lagergren, 1898), based on solid capacity, is generally expressed as:

$$\frac{dq_t}{dt} = k_1 (q_e - q_t) \qquad\qquad (6\text{-}1)$$

Integrating this for the boundary conditions $t = 0$ to $t = t$ and $q_t = q_t$, equation (6-1) may be rearranged for linearised data plotting as shown below:

$$\log(q_e - q_t) = \log(q_e) - \frac{k_1}{2.303} t \qquad\qquad (6\text{-}2)$$

where q_t and q_e are the amount of solute sorbed per mass of sorbent (mg g^{-1}) at any time and equilibrium, respectively, and k_1 is the rate constant of first-order sorption (min^{-1}). The straight-line plot of $\log(q_e - q_t)$ against t gives $\log(q_e)$ as slope and intercept equal to $k_1/2.303$. Hence the amount of solute sorbed per gram of sorbent at equilibrium (q_e) and the first-order sorption rate constant (k_1) can be evaluated from the slope and the intercept.

6.3.1.2 *Pseudo second-order kinetic model*

The pseudo-second-order equation (Ho and MacKay, 1999; Azizian, 2004; Iftikhar et al., 2009) is also based on the sorption capacity of the solid phase. If the rate of sorption is a second-order mechanism, the pseudo second-order kinetic rate equation is expressed as:

$$\frac{dq_t}{dt} = k_2 (q_e - q_t)^2 \qquad\qquad (6\text{-}3)$$

Integrating eq. (6-3) for the boundary conditions $t = 0$ and $t = t$ and $q = 0$ to $q = q$, gives:

$$\frac{1}{(q_e - q_t)} = \frac{1}{q_e} + k_2 t \qquad\qquad (6\text{-}4)$$

where k_2 is the second-order rate constant for sorption. Equation (4) can be rearranged to obtain a linear form as:

$$\frac{t}{q_t} = \frac{1}{k_2 q_e^2} + \frac{1}{q_e} t \qquad\qquad (6\text{-}5)$$

The initial sorption rate is defined as (Pamukoglu and Kargi, 2006; Zhou et al., 2009):

$$h = k_2 q_e^2 \qquad\qquad (6\text{-}6)$$

The plot of t/q_t against t gives a straight line with intercept of $1/k_2 q_e^2$ and intercept of $1/q_e$. Hence the amount of solute sorbed per gram of sorbent at equilibrium (q_e) and the sorption rate constant (k_2) can be evaluated from the slope and the intercept.

6.3.1.3   *Intraparticle diffusion model*
During the batch mode of sorption, there is a possibility of transport of Cu(II) into the pores of the coconut shell and *Moringa oleifera* seeds, which is often the rate controlling step. The most commonly used technique for identifying the mechanism involved in the adsorption process is by fitting an intraparticle diffusion plot. The intraparticle diffusion model is given by the equation (Ncibi et al., 2008; Kalavathy et al., 2005):

$$q_t = k_{id} t^{0.5} \qquad\qquad (6\text{-}7)$$

where, k_{id} is the intraparticle diffusion rate constant (mg g^{-1} min$^{-0.5}$). The plot of q_t vs $t^{0.5}$ represents the different stages of adsorption. If this plot represents multi-linearity in its shape, such behaviour characterises the involvement of two or more steps in the overall sorption process (Liu and Wang, 2008). The initial curved portion relates to the boundary layer diffusion (film diffusion) and the latter linear portion represents the intraparticle diffusion. The slope of the second linear portion of the plot has been defined to yield the intraparticle diffusion parameter k_{id} (mg g^{-1} min$^{-0.5}$). On the other hand, the intercept of the plot reflects the boundary layer effect. The bigger the intercept, the greater is the contribution of the surface sorption to the rate-controlling step.

6.3.1.4   *The Elovich kinetic equation*
The Elovich kinetic equation was first used to describe the rate of adsorption of carbon monoxide on manganese dioxide (Ho, 2006; Qiu et al., 2009). The Elovich kinetic equation is generally expressed as (Low, 1960; Chein and Clayton, 1980; Augustine et al., 2007):

$$\frac{dq_t}{dt} = \alpha e^{-aq_t} \qquad\qquad (6\text{-}8)$$

where q_t is the sorption capacity at time t (mg g^{-1}), α is the initial adsorption rate (mg g^{-1}min^{-1}) and a is desorption constant (mg g^{-1} min^{-1}) during any experiment. Equ. (6-7) is arranged to a linear form as:

$$q_t = \left(\frac{2.3}{\alpha}\right)\log(t + t_0) - \left(\frac{2.3}{\alpha}\right)\log t_0 \qquad\qquad (6\text{-}9)$$

Where

$$t_0 = \frac{1}{\alpha} a \qquad\qquad (6\text{-}10)$$

To simplify equation (6-8), it was assumed that $a\alpha t \gg 1$ (Chein and Clayton, 1980). By integrating and applying boundary conditions $q_t = 0$ at $t = 0$ and $q_t = q_t$ at $t = t$ [34,37], equation (7) then becomes:

$$q_t = \alpha \ln(a\alpha) + \alpha \ln(t) \qquad\qquad (11)$$

A plot of q_t versus ln t gives a straight line from which α and a can be obtained from the slope and intercept.

6.3.1.5 *Mass transfer analysis*

The uptake of pollutant species from liquid phase (sorbate) to solid phase (sorbent) is carried out by transfer of mass from the former to the latter. The overall sorption process is assumed to occur in three steps (Hasan et al., 2008):

1. Mass transfer of sorbate from the aqueous phase on to the solid surface,
2. Sorption of solute on to the surface sites,
3. Internal diffusion of solute via either a pore diffusion model or homogeneous solid phase diffusion model.

Mass transfer analysis for Cu(II) sorption onto coconut shell and *Moringa oleifera* seeds was carried out using the equation as proposed by McKay et al. (1981) and also applied by Naiya et al. (2009); Hasan et al. (2008):

$$\ln\left(\frac{C_t}{C_0} - \frac{1}{1+mK}\right) = \ln\left(\frac{mK}{1+mK}\right) - \ln\left(\frac{1+mK}{mK}\right)\beta_l S_s t \tag{6-12}$$

where m is the mass of biosorbent per unit volume, K is the constant obtained by multiplying q_{max} and K_L (Langmuir constants), β_l (cm s^{-1}) is the mass transfer coefficient and S_s (cm^{-1}) is the outer specific surface of the biosorbent particles per unit volume of the particle free slurry. Therefore, a plot of $\ln[C_t/C_0 - 1/(1 + mK)]$ versus t will yield a straight line of intercept $\ln[mK/(1 + mK)]$and slope $-[(1 + mK)/mK]\beta_l S_s$, from which the surface mass transfer coefficient can be obtained.

6.3.2 Calculations

The metal uptake per gram sorbent, the percentage removal and the desorption efficiency were calculated according to the following equations:

$$q_e = \frac{(C_0 - C_e) \times V}{m} \tag{6-13}$$

$$revoval(\%) = \frac{(C_0 - C_e)}{C_0} \times 100 \tag{6-14}$$

where q_e is the equilibrium adsorption capacity (mg g^{-1}),C_0 is the initial concentration (mg L^{-1}) of metal ions in solution, C_e is the equilibrium concentration (mg L^{-1}), of metal ions in solution, V is the volume of aqueous solution (L) and m is the dry weight of the adsorbent per unit volume of particles free slurry (g L^{-1}). *Removal (%)* The outer specific surface of the biosorbent particles per unit volume of the particle free slurry was calculated from the following equation (MacKay et al., 1981):

$$S_s = \frac{6m}{d_p \rho_p (1 - \varepsilon_p)} \tag{6-15}$$

where d_p is the diameter (cm), ρ_p is the density (g cm^{-3}) and ε_p is the porosity of the biosorbent particles

6.4   Results

6.4.1   Kinetics of Cu(II) biosorption onto coconut shell and *Moringa oleifera* seeds

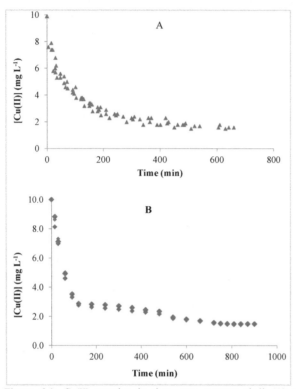

Figure 6-1: Cu(II) sorption in time onto coconut shell (A) and *Moringa oleifera* seeds (B). Experimental conditions: initial Cu(II) concentration = 10 mg L^{-1}, sorbent concentration = 20 g L^{-1}, particle size = 0.5 - 0.8 mm, mixing rate = 100 rpm, temperature = 30 ± 0.2 ^{0}C, pH = 7.0 ± 0.2.

Figure 6-1 shows the removal of Cu(II) in time by 20 g L^{-1} coconut shell and *Moringa oleifera* seeds. In the case of coconut shell, it is apparent that sorption is the dominant process occurring in this experiment. The Cu(II) removal increases rapidly within the first 3 hours, reaching 70%, and where a maximum removal of 85% of the Cu(II) in solution was achieved. The process lasted for 660 minutes (Figure 6-1A).

In the case of the *Moringa oleifera* seeds, the extent of sorption increased rapidly in the initial stages (up to 120 minutes) but became slow in the later stages till attainment of equilibrium in 900 minutes (Figure 6-1B).

6.4.2   Sorption kinetics modelling
6.4.2.1   *Pseudo first-order kinetics modelling*
Figure 6-2 shows the pseudo first-order kinetics modelling of Cu(II) biosorption onto coconut shell and *Moringa oleifera* seeds. The figures show a deviation from the straight line in the first 30 min of sorption. The values of the model parameters and the correlation coefficients are presented in Table 6-2. The equilibrium rate constant

of the pseudo first-order sorption (k_1) for coconut shell and *Moringa oleifera* seeds are
2.0223×10^{-4} min^{-1} and 1.842×10^{-3} min^{-1}, respectively. From Table 6-2, it is
observed that *Moringa oleifera* seeds ($R^2 = 0.9659$) gave a better fit than coconut
shell ($R^2 = 0.8189$).

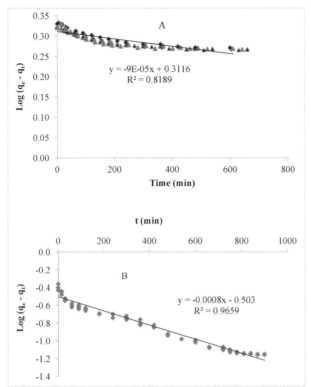

Figure 6-2: Pseudo first-order plots for Cu(II) biosorption unto coconut shell (A) and *Moringa oleifera*
seeds (B). Experimental conditions: initial Cu(II) concentration = 10 mg L^{-1}, sorbent concentration =
20 g L^{-1}, particle size = 0.5 - 0.8 mm, mixing rate = 100 rpm, temperature = 30 ± 0.2 ^{0}C, pH = 7.0 ±
0.2.

6.4.2.2 *Pseudo second-order kinetics model*

The equilibrium rate constants of the pseudo second-order kinetics for Cu(II) sorption
were determined from the plot of t/q_t versus time as shown in Figure 3. The initial
sorption rate (h), the rate constant (k_2) and the correlation coefficient (R^2) of this
model for the coconut shell (Figure 6-3A) and *Moringa oleifera* seeds plots (Figure 6-
3B) were calculated and presented in Table 6-2. The results show very good
correlation coefficients for Cu(II) removal by both biosorbents. However, the coconut
shell gave a slightly better fit ($R^2 =0.9971$) than the *Moringa oleifera* seeds ($R^2 =
0.9958$).

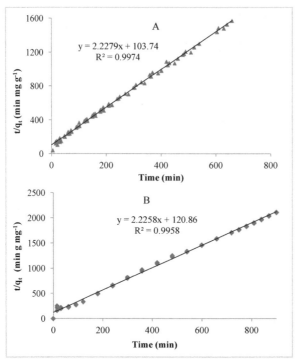

Figure 6-3: Pseudo second-order plots for Cu(II) biosorption unto coconut shell (A) and *Moringa oleifera* seeds (B). Experimental conditions:　initial Cu(II) concentration = 10 mg L^{-1}, sorbent concentration = 20 g L^{-1}, particle size = 0.5 - 0.8 mm, mixing rate = 100 rpm, temperature = 30 ± 0.2 ^{0}C, pH = 7.0 ± 0.2.

6.4.2.3　　　　Intraparticle diffusion model

Figure 6-4 presents the intraparticle diffusion plot for Cu(II) sorption onto coconut shell and *Moringa oleifera* seeds. The plots show two phases: the initial part of the curve is attributed to the boundary layer diffusion effect (i.e. external film resistance), while the final part indicates the intraparticle diffusion. The intraparticle diffusion rate constants were calculated from the slope of the second phase of the plots and are presented in Table 6-2. The intercept of these plots is proportional to the thickness of the boundary layer, and was calculated for Cu(II) sorption onto coconut shell (Figure 6-4A) and *Moringa oleifera* seeds (Figure 6-4B) as 0.273 mg g^{-1} and 0.294 mg g^{-1}, respectively.

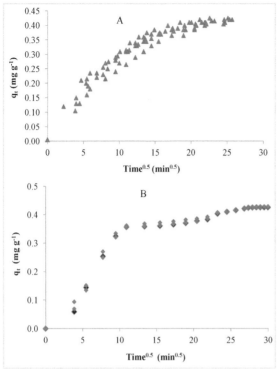

Figure 6-4: Intraparticle diffusion plots for Cu(II) biosorption unto coconut shell (A) and *Moringa oleifera* seeds (B). Experimental conditions: initial Cu(II) concentration = 10 mg L^{-1}, sorbent concentration = 20 g L^{-1}, particle size = 0.5 - 0.8 mm, mixing rate = 100 rpm, temperature = 30 ± 0.2 ^{0}C, pH = 7.0 ± 0.2.

6.4.2.4   The Elovich kinetic modelling

Figure 6-5 shows the Elovich equation plot for Cu(II) sorption onto coconut shell and *Moringa oleifera* seeds. Table 6-2 shows the Elovich equation constants, α (the initial sorption rate) and *a* (the desorption capacity), obtained from the slope and intercept of the plots for coconut shell (Figure 6-5A) and *Moringa oleifera* seeds (Figure 6-5B). The correlation coefficients obtained from the plots for the two biosorbents (Table 6-2) indicate that the model fits the experimental data fairly well. Coconut shell gave a better fit (R^2 = 0.9850) compared to the R^2 = 0.9415 for *Moringa oleifera* seeds. There exists an inverse relationship between α and *a* (Table 6-2), since sorption and desorption are interrelated in surface transport.

6.4.2.5   Mass transfer modelling

Figure 6-6 shows the mass transfer plots for Cu(II) sorption onto coconut shell and *Moringa oleifera* seeds. The values of the mass transfer coefficient β_l calculated from the slope and the intercept of the plots for coconut shell (Figure 6-6A) and *Moringa oleifera* seeds (Figure 6-6B) are 1.2106 × 10^{-3} cm s^{-1} and 8.965 × 10^{-2} cm s^{-1}, respectively (Table 6-2). The correlation coefficients obtained from the plots indicate a slightly better fit for *Moringa oleifera* seeds data (R^2 = 0.9752) than for coconut shell (R^2 = 0.9345).

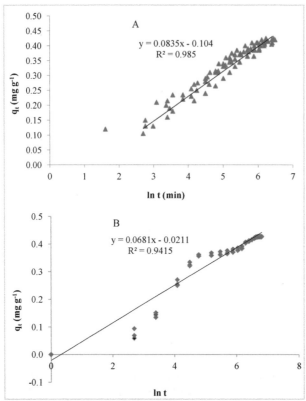

Figure 6-5: Elovich plots for Cu(II) biosorption unto coconut shell (A) and *Moringa oleifera* seeds (B). Experimental conditions: initial Cu(II) concentration = 10 mg L^{-1}, sorbent concentration = 20 g L^{-1}, particle size = 0.5 - 0.8 mm, mixing rate = 100 rpm, temperature = 30 ± 0.2 ^{0}C, pH = 7.0 ± 0.2.

Table 6-2: Summary of the kinetics model parameters for Cu(II) sorption onto the coconut shell and
the *Moringa oleifera* seeds

Kinetics parameter	Biosorbent	
	Coconut shell	*Moringa oleifera* seeds
Pseudo first-order model		
k_1 (min^{-1})	2.023×10^{-4}	1.842×10^{-3}
R^2	0.8189	0.9659
Pseudo second-order model		
k_2 (g mg^{-1} min^{-1})	4.785×10^{-2}	4.126×10^{-2}
h (mg g^{-1} min^{-1})	9.6395×10^{-3}	8.3292×10^{-3}
R^2	0.9974	0.9958
Intraparticle diffusion model		
k_{id1} (mg g^{-1} min$^{-0.5}$)	2.62×10^{-2}	3.55×10^{-2}
k_{id2} (mg g^{-1} min$^{-0.5}$)	5.90×10^{-3}	4.50×10^{-3}
R^2	0.9336	0.9440
Elovich model		
α (mg g^{-1} min^{-1})	8.35×10^{-2}	6.81×10^{-2}
a (mg g^{-1} min^{-1})	3.4464	10.7719
R^2	0.9850	0.9415
Mass transfer model		
β_1 (cm s^{-1})	1.2106×10^{-3}	8.965×10^{-4}
R^2	0.9345	0.9752

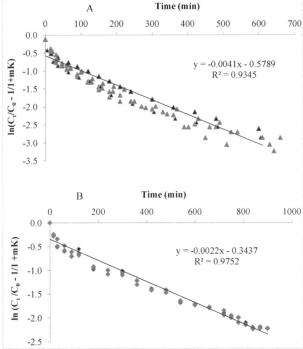

Figure 6-6: Mass transfer plots for Cu(II) biosorption unto coconut shell (A) and *Moringa oleifera*
seeds (B). Experimental conditions: initial Cu(II) concentration = 10 mg L^{-1}, sorbent concentration =
20 g L^{-1}, particle size = 0.5 - 0.8 mm, mixing rate = 100 rpm, temperature = 30 ± 0.2 ^{0}C, pH = 7.0 ±
0.2.

6.5   Discussion

6.5.1   Kinetics of Cu(II) biosorption onto coconut shell and *Moringa oleifera* seeds

The results (Figure 6-1) indicate the potential of using the studied biosorbents to treat gold mine wastewater having a Cu(II) concentration of up to 10 mg L^{-1}. The *Moringa oleifera* seeds removed 74% of the Cu(II) in 180 minutes (Figure 6-1B), compared to the 70% removal by the coconut shell in the same time (Figure 6-1A). However, coconut shell gave the highest overall removal percentage (85% in 600 minutes), whereas *Moringa oleifera* seeds achieved the same removal in 900 minutes only (Figure 6-1B). Therefore coconut shell has the lower equilibrium time (600 minutes) for the sorption of Cu(II), making it a better choice for the removal of Cu(II) from gold mine wastewater. Furthermore, the reduction in the Cu(II) concentration in the wastewater from 10 mg L^{-1} to less than 2 mg L^{-1} (Figure 6-1) indicates the ability of the biosorbents to remove Cu(II) from the wastewater to meet the Ghana EPA discharge limit of 2.0 mg L^{-1} (Ghana EPA, 2010).

6.5.2   Sorption kinetics modelling

The two chemical reaction rate models, namely, the first- and second-order equation, used in fitting the experimental data gave an indication of the reaction pathway. From the two models, the pseudo second-order equation fits the experimental data better with a correlation coefficient close to unity (Table 6-2). On the basis of this parameter fitting, the biosorption of Cu(II) onto the coconut shell or the *Moringa oleifera* seeds is a second order reaction, and not a first-order one. The deviation from the straight line in the first 30 min of sorption, observed for the pseudo first-order model, is attributed to the fast initial uptake of Cu(II) onto the biosorbents prior to the slow biosorption in the pores. A review of the literature (Table 6-3) shows that the biosorption of Cu(II) onto natural and agricultural materials usually follows the pseudo second-order kinetics. The second-order rate constant obtained in those researches varies between 0.001-0.86 g mg^{-1} min^{-1} (Table 6-3).

Table 6-3: Data on pseudo second-order kinetics rate constant for Cu(II) sorption using different biosorbents

Biosorbent	Temp (K)	k$_2$ (g mg^{-1} min^{-1})	R^2	Reference
Zeolite NaX	308	0.021	0.984	Svilovic et at. (2009)
Cassava	308	0.860	0.998	Horsfall et al. (2006)
Acid modified cassava	301	0.656	-	Augustine et al. (2007)
Low grade rock phosphate (francolite)	303	0.008	0.946	Prasad et al. (2008)
Pre-treated powdered waste sludge	298	0.001	0.990	Pamukoglu and Fargi (2006)
Marine algae	298	0.090	0.998	Abdelwahab (2007)
Green algae activated carbon	298	0.020	0.997	Abdelwahab (2007)
Chestnut shell	303	0.217	0.999	Yao et al. (2010)
Pine cone powder	307	0.082	0.997	Ofomaja et al. (2010)
Mushroom biomass	293	0.162	0.998	Ertugay et al. (2010)
Untreated coffee husk	298	0.661	0.999	Oliveira et al. (2008)
Coconut shell	303	0.048	0.997	This study
Moringa oleifera seeds	303	0.041	0.996	This study

The Elovich equation described Cu(II) sorption onto coconut shell better than *Moringa oleifera* seeds. In the case of the *Moringa oleifera* seeds, the plot did not represent a straight line; though the correlation coefficient is fairly good. On the other hand, the sorption of Cu(II) by coconut shell was described by the Elovich model (R^2

Chapter 6: Kinetics modelling of Cu (II) biosorption onto coconut shell and *moringa oleifera* seeds
from tropical regions
113

= 0.985), indicating the applicability of this model to the process. The inverse relationship observed between the initial sorption rate constant and the desorption constant (Table 6-2), confirmed the interrelationship between sorption and desorption in surface transport. The inverse relationship between the two constants was also confirmed in the study by Augustine et al. (2007) and Razmovski and Scidan (2008). The Elovich model predicts that a diffusion step controls the sorption of Cu(II) by coconut shell.

The prediction of the rate limiting step is an important factor to be considered in the adsorption process. It is governed by the adsorption mechanism, which is generally used for design purposes. The overall sorption rate is usually controlled by the slowest step, which could be either the film diffusion or the intraparticle (pore) diffusion (Allen et al., 1989; Chaithanya and Yedla, 2010).

The deviation of the plots for Cu(II) sorption onto the coconut shell (Figure 6-4A) and *Moringa oleifera* seeds (Figure 6-4B) from the origin may be attributed to the differences in the mass transfer rate in the initial and final stages of sorption. Such a deviation of a straight line from the origin indicates that the film diffusion is also important in the sorption process. The intraparticle diffusion rate (K_{id2}) obtained for Cu(II) sorption by the two biosorbents (Table 6-2) indicates that the rate limiting step is the intraparticle diffusion and not the film diffusion. Hence the intraparticle diffusion controls the biosorption rate in this study. Nonetheless, the external mass transfer resistance can not be neglected, though this is only significant in the initial stage of sorption.

Finally, the mass transfer coefficients obtained for Cu(II) sorption onto coconut shell and *Moringa oleifera* seeds (Table 6-2) revealed that the rate of mass transfer from the bulk solution to the biosorbent surfaces was quite fast, and therefore cannot be considered as the rate limiting step. The deviation of some points of the plots from linearity is an indication of the varying extent of mass transfer in the initial and final stages of the sorption process (Gupta et al., 1988).

6.6 Conclusions

Batch sorption studies for the removal of Cu(II) from gold mine wastewater have been carried out using coconut shell and *Moringa oleifera* seeds as low cost, locally available biosorbents. The present work illustrates the importance of chemical reaction kinetics, adsorption diffusion and mass transfer models to the sorption of Cu(II) onto the studied biosorbents. The rate of sorption of Cu(II) was rapid in the initial 180 min (coconut shell) and 90 min (*Moringa oleifera* seeds), and then continued slowly until equilibrium was attained. The equilibrium time for biosorption of Cu(II) by coconut is 660 minutes, while equilibrium was attained after 900 minutes in the case of *Moringa oleifera* seeds. The kinetic data for the Cu(II) sorption were fitted to five kinetics models, namely, the pseudo first- and second-order, intraparticle diffusion, Elovich and mass transfer model. The experimental data obtained from the rate kinetics were better described by the pseudo second-order model than the first-order model, as evident from the correlation coefficient (R^2) for both biosorbents. Mass transfer studies showed that the mass transfer rate from the bulk solution to the surface of the biosorbent particles was fast, and was not the rate limiting step. Intraparticle diffusion was found to be the rate limiting step, and therefore controls the sorption process in this study. However, film diffusion can not be ignored, especially, at the initial stage of sorption due to the dual nature of the intraparticle diffusion plots for both biosorbents.

6.7   Acknowledgement

The authors acknowledge funding from the Netherlands Fellowship Programme (NFP), Staff Development and Postgraduate Scholarship Scheme (Kumasi Polytechnic, Ghana) and the UNESCO-IHE Partner Research Fund (UPaRF). We further acknowledge cooperation with AngloGold Ashanti (Obuasi, Ghana).

6.8   References

Abdelwahab, O., 2007. Kinetic and isotherm studies of copper (II) removal from wastewater using various adsorbents. Egyptian Journal of Aquatic Research 33 , 125-143.

Acheampong, M.A., Meulepas, R.J., Lens, P.N., 2010. Removal of heavy metals and cyanide from gold mine wastewater. Journal of Chemical Technology & Biotechnology 85, 590-613.

Acheampong, M.A., Pereira, J.P.C., Meulepas, R.J.W., Lens, P.N.L., 2011. Biosorption of Cu(II) onto agricultural materials from tropical regions. Journal of Chemical Technology & Biotechnology 86, 1184-1194.

Ahmady-Asbchin, S., Andrès, Y., Gérente, C., Cloirec, P.L., 2008. Biosorption of Cu(II) from aqueous solution by Fucus serratus: Surface characterization and sorption mechanisms. Bioresource Technology 99, 6150-6155.

Allen, S.J., McKay, G., Khader, K.Y.H., 1989. Intraparticle diffusion of a basic dye during adsorption onto sphagnum peat. Environmental Pollution 56, 39-50.

Anirudhan, T.S., Radhakrishnan, P.G., 2008. Thermodynamics and kinetics of adsorption of Cu(II) from aqueous solutions onto a new cation exchanger derived from tamarind fruit shell. The Journal of Chemical Thermodynamics 40, 702-709.

Augustine, A.A., Orike, B.D., Edidiong, A.D., 2007. Adsorption kinetics and modelling of Cu(II) ion sorption from aqueous solution by mercaptoacetic acid modified cassava (Manihot Sculenta Cranz) waste. Electronic Journal of Environmental, Agricultural and Food Chemistry 6 (4), 2221-2234.

Azizian, S., 2004. Kinetic models of sorption: a theoretical analysis. Journal of Colloid and Interface Science 276, 47-52.

Beolchini, F., Pagnanelli, F., Toro, L., Vegliò, F., 2003. Biosorption of copper by Sphaerotilus natans immobilised in polysulfone matrix: equilibrium and kinetic analysis. Hydrometallurgy 70, 101-112.

Chein, S.H., Clayton, W.R., 1980. Application of Elovich equation to the kinetics of phosphate release and sorption on soil, Soil Science Society of America Journal 44, 265.

Farooq, U., Kozinski, J.A., Khan, M.A., Athar, M., 2010. Biosorption of heavy metal ions using wheat based biosorbents - A review of the recent literature. Bioresource Technology 101, 5043-5053.

Febrianto, J., Kosasih, A.N., Sunarso, J., Ju, Y.-H., Indraswati, N., Ismadji, S., 2009. Equilibrium and kinetic studies in adsorption of heavy metals using biosorbent: A summary of recent studies. Journal of Hazardous Materials 162, 616-645.

Gadd, G.M., 2009. Biosorption: critical review of scientific rationale, environmental importance and significance for pollution treatment. Journal of Chemical Technology & Biotechnology 84, 13-28.

Chaithanya, T.K., Yedla, S., 2010. Adsorption of hexa-valent chromium using treated wood charcoal – elucidation of rate-limiting process. Environmental Technology 31, 1495-1505.

Ghana EPA, 2010. Environmental performance rating and disclosure: report on the performance of mining and manufacturing companies. Environmental Protection Agency, Accra, Ghana.

Chojnacka, K., 2006. Biosorption of Cr (III) ions by waste straw and grass: a systematic characterization of new biosorbents. Polish Journal of Environmental Studies 15 (6), 845-852.

Ertugay, N., Bayhan, Y.K., 2010. The removal of copper (II) ion by using mushroom biomass (Agaricus bisporus) and kinetic modelling. Desalination 255, 137-142.

Goel, J., Kadirvelu, K., Rajagopal, C., 2004. Competitive sorption of Cu(II), Pb(II) and Hg(II) ions

from aqueous solution using coconut shell-based activated carbon. Adsorption Science and
Technology 22, 257-273.

Gulnaz, O., Saygideger, S., Kusvuran, E., 2005. Study of Cu(II) biosorption by dried activated sludge:
effect of physico-chemical environment and kinetics study. Journal of Hazardous Materials
120, 193-200.

Gupta, G.S., Prasad, G., Singh, V.N., 1988. Removal of color from wastewater by sorption for water
reuse. Journal of Environmental Science and Health - Part A Environmental Science and
Engineering 23, 205-217.

Hameed, B.H., El-Khaiary, M.I., 2008. Batch removal of malachite green from aqueous solutions by
adsorption on oil palm trunk fibre: Equilibrium isotherms and kinetic studies. Journal of
Hazardous Materials 154, 237-244.

Hasan, S.H., Singh, K.K., Prakash, O., Talat, M., Ho, Y.S., 2008. Removal of Cr(VI) from aqueous
solutions using agricultural waste "maize bran'. Journal of Hazardous Materials 152, 356-365.

Ho, Y.-S., 2006. Review of second-order models for adsorption systems. Journal of Hazardous
Materials 136, 681-689.

Ho, Y.S., McKay, G., 1999. Pseudo-second order model for sorption processes. Process Biochemistry
34, 451-465.

Ho, Y.S., Wang, C.C., 2004. Pseudo-isotherms for the sorption of cadmium ion onto tree fern. Process
Biochemistry 39, 761-765.

Horsfall, J.M., Abia, A.A., Spiff, A.I., 2006. Kinetic studies on the adsorption of Cd^{2+}, Cu^{2+} and Zn^{2+}
ions from aqueous solutions by cassava (Manihot sculenta Cranz) tuber bark waste.
Bioresource Technology 97, 283-291.

Iftikhar, A.R., Bhatti, H.N., Hanif, M.A., Nadeem, R., 2009. Kinetic and thermodynamic aspects of
Cu(II) and Cr(III) removal from aqueous solutions using rose waste biomass. Journal of
Hazardous Materials 161, 941-947.

Kalavathy, M.H., Karthikeyan, T., Rajgopal, S., Miranda, L.R., 2005. Kinetic and isotherm studies of
Cu(II) adsorption onto H_3PO_4-activated rubber wood sawdust. Journal of Colloid and Interface
Science 292, 354-362.

Lagergren, S., 1898. About the theory of so-called adsorption of soluble substances. Kungliga Svenska
Vetenskapsa-kademiens. Handlingar 24(4), 1-34.

Liu, Y., Wang, Z.-W., 2008. Uncertainty of preset-order kinetic equations in description of biosorption
data. Bioresource Technology 99, 3309-3312.

McKay, G., Otterburn, M.S., Sweeney, A.G., 1981. Surface mass transfer processes during colour
removal from effluent using silica. Water Research 15, 327-331.

Naiya, T.K., Bhattacharya, A.K., Das, S.K., 2009. Adsorption of Cd(II) and Pb(II) from aqueous
solutions on activated alumina. Journal of Colloid and Interface Science 333, 14-26.

Ncibi, M.C., Mahjoub, B., Seffen, M., 2008. Investigation of the sorption mechanisms of metal-
complexed dye onto *Posidonia oceanica* (L.) fibres through kinetic modelling analysis.
Bioresource Technology 99, 5582-5589.

Ngah, W.S.W., Hanafiah, M.A.K.M., 2008. Biosorption of copper ions from dilute aqueous solutions
on base treatedrubber (Hevea brasiliensis) leaves powder: kinetics, isotherm, and biosorption
mechanisms. Journal of Environmental Sciences 20, 1168-1176.

Ofomaja, A.E., Naidoo, E.B., Modise, S.J., 2010. Dynamic studies and pseudo-second order modeling
of copper(II) biosorption onto pine cone powder. Desalination 251, 112-122.

Oliveira, W.E., Franca, A.S., Oliveira, L.S., Rocha, S.D., 2008. Untreated coffee husks as biosorbents
for the removal of heavy metals from aqueous solutions. Journal of Hazardous Materials 152,
1073-1081.

Pamukoglu, M.Y., Kargi, F., 2006. Batch kinetics and isotherms for biosorption of copper(II) ions onto
pre-treated powdered waste sludge (PWS). Journal of Hazardous Materials 138, 479-484.

Prasad, M., Xu, H.-y., Saxena, S., 2008. Multi-component sorption of Pb(II), Cu(II) and Zn(II) onto
low-cost mineral adsorbent. Journal of Hazardous Materials 154, 221-229.

Qiu, H., Pan, B-C., Zhang, Q-J., Zhang, W-M., Zhang, Q-X., 2009. Critical review in adsorption
kinetic models. Journal of Zhejiang University - Science A 10, 716-724.

Razmovski, R., Sciban, M., 2008. Biosorption of Cr(VI) and Cu(II) by waste tea fungal biomass.
Ecological Engineering 34, 179-186.

Rudzinski, W., Plazinski, W., 2007. Theoretical description of the kinetics of solute adsorption at
heterogeneous solid/solution interfaces: On the possibility of distinguishing between the
diffusional and the surface reaction kinetics models. Applied Surface Science 253, 5827-5840.

Sajidu, S.M.I., Persson, I., Masamba, W.R.L., Henry, E.M.T., 2008. Mechanisms for biosorption of chromium(III), copper(II) and mercury(II) using water extracts of Moringa oleifera seed powder. African Journal of Biotechnology 7, 800-804.

Sengil, I.A., Özacar, M., Türkmenler, H., 2009. Kinetic and isotherm studies of Cu(II) biosorption onto valonia tannin resin. Journal of Hazardous Materials 162, 1046-1052.

Svilovic, S., Rusic, D., Stipisic, R., 2009. Modeling batch kinetics of copper ions sorption using synthetic zeolite NaX. Journal of Hazardous Materials 170, 941-947.

Vilar, V.J.P., Botelho, C.M.S., Boaventura, R.A.R., 2007. Methylene blue adsorption by algal biomass based materials: Biosorbents characterization and process behaviour. Journal of Hazardous Materials 147, 120-132.

Volesky, B., 2001. Detoxification of metal-bearing effluents: biosorption for the next century. Hydrometallurgy 59, 203-216.

Yao, Z.Y., Qi, J.H., Wang, L.H., 2010. Equilibrium, kinetic and thermodynamic studies on the biosorption of Cu(II) onto chestnut shell. Journal of Hazardous Materials 174, 137-143.

Zhou, L., Wang, Y., Liu, Z., Huang, Q., 2009. Characteristics of equilibrium, kinetics studies for adsorption of Hg(II), Cu(II), and Ni(II) ions by thiourea-modified magnetic chitosan microspheres. Journal of Hazardous Materials 161, 995-1002.

Chapter 7

7 Removal of Cu(II) by biosorption onto coconut shell in fixed-bed column systems

This has been presented and published as:

Acheampong, M.A., Lens, P.N.L., 2012.   Effect of operating parameters on Cu(II) biosorption onto coconut shell in a fixed-bed column. In: ***Proceedings of the 2013 International Conference on Pollution and Treatment Technology***, Hainan Island, China (2 - 4 January 2012).

Acheampong, M.A., Pakshirajan, K., Annachhatre, A.P., Lens, P.N.L., 2013. Removal of Cu(II) by biosorption onto coconut shell in fixed-bed column systems. Journal of Industrial and Engineering Chemistry 19, 841-848.

Abstract

The performance of a fixed-bed column packed with coconut shell for the biosorption of Cu(II) ions was evaluated using column breakthrough data at different flow rates, bed-depths and initial Cu(II) concentrations. The Bed Depth Service Time (BDST), Yoon-Nelson, Thomas and Clark models were used to evaluate the characteristic design parameters of the column. The Cu(II) biosorption column had the best performance at 10 mg L^{-1} inlet Cu(II) concentration, 10 mL min^{-1} flow rate and 20 cm bed depth. Under these optimum conditions, the service time to breakthrough was about 60 h, after which the Cu(II) concentration in the effluent exceeded the 1 mg L^{-1} discharge limit set by the Ghana Environmental Protection Agency (EPA). The equilibrium uptake of Cu(II) amounted to 7.25 mg g^{-1}, which is 14.5 times higher than the value obtained in a batch study with the same material for the same initial Cu(II) concentration (10 mg L^{-1}). The BDST model fitted well the experimental data in the 10% and 50% regions of the breakthrough curve. The Yoon-Nelson model predicted well the time required for 50% breakthrough (τ) at all conditions examined. The simulation of the whole breakthrough curve was successful with the Yoon-Nelson model, but the breakthrough curve was best predicted by the Clark model. The design of a fixed bed column for Cu(II) removal from wastewater by biosorption onto coconut shell can be done based on these models.

Keywords: Adsorption; Biosorption; Breakthrough curve; Copper; Fixed-bed column; Wastewater

7.1   Introduction

In recent years increasing concern about the effect of toxic metals in the environment has resulted in more stringent environmental regulations for industrial operations that discharge metal-bearing effluents. Copper is essential for good health, but it is also potentially toxic to humans when taken in excessive amounts (Stylianou et al., 2007; Lu et al., 2009; Shi et al., 2009). Therefore, limits have been set for copper in wastewater prior to discharge into the environment. For example, the current Ghana Environmental Protection Agency (EPA) standard for copper discharge is 1.0 mg L^{-1} (Ghana EPA, 2010). To meet the discharge standards, copper has to be removed from metal rich wastewaters, such as those from the mining industry.

Existing physico-chemical treatment technologies require continuous input of chemicals, making them expensive, environmentally unfriendly and ineffective for low strength wastewaters (Tung et al., 2002; Han et al., 2006a; Amarasinghe and Williams, 2007; Apiratikul and Pavasant, 2008; Fagundes-Klen et al., 2010). Research in recent years has indicated that some natural biomaterials, including agricultural products and by-products, can accumulate high concentrations of heavy metals (Qaiser et al., 2009; Wu et al., 2010; Nayek et al., 2010). The process by which these biomaterials sequester heavy metals from aqueous solutions is commonly referred to as biosorption and involves one or more combinations of mechanisms such as complexation, microprecipitation and ion exchange (Vijayaraghavan and Prabu, 2006). Low cost and relatively high heavy metals removal efficiency from dilute solutions are major advantages (Naja and Volesky, 2006). As compared to the many biosorbents described in the literature, coconut shell, an abundant agricultural product, has shown good ability to remove copper from wastewater. Coconut shell can be considered as a low-cost biosorbent for heavy metal removal from wastewaters (Acheampong et al., 2011). Studies with this biosorbent so far have, however, been restricted to batch systems only. Further investigations are warranted to establish its potential under a continuous mode in a fixed bed column.

In order to validate the biosorption data obtained under batch conditions (Acheampong et al., 2011) (Chapter 4), evaluation of sorption performance in a continuously operated column is necessary because the sorbent uptake capacity is more efficiently utilised than in a completely mixed system and also the contact time required to attain equilibrium is different under column operation mode (Chandra-Sekhar et al., 2003; Vilar et al., 2008). For column operation, the adsorbent is continuously in contact with fresh wastewater and consequently, the concentration in the solution in contact with a given layer of the biosorbent in a column changes very slowly. A fixed-bed column is simple to operate and economically valuable for wastewater treatment (Singh et al., 2009; Sousa et al., 2010; Ahmad and Hmeed 2010). Experiments using a laboratory-scale fixed-bed column of relatively large volume yield performance data that can be used to design a larger pilot and industrial scale plant with a high degree of accuracy (Chandra-Sekhar et al., 2003).

This study was aimed at continuous removal of Cu(II) by coconut shell in a fixed-bed column to determine the optimum operating conditions. Further, the breakthrough data obtained for each operating condition were fitted to adsorption models to enable appropriate design and scale up of the sorption column. The coconut shell investigated in this study has previously been shown to be effective in sorbing Cu(II) from aqueous solutions at pH 7 ± 0.2 and in the particle size range 0.5-1.4 mm (Acheampong et al., 2011) (Chapter 4). These conditions were therefore maintained in the column experiments as well for examining the influence of various operating parameters on continuous Cu(II) sorption by coconut shell.

7.2   Materials and methods

7.2.1   Biosorbent preparation

The coconut shell used in this study was obtained from Kumasi (Ghana) and its Cu(II) biosorption potential under batch conditions was described in detail by Acheampong et al. (2011). The material was washed with distilled water and dried at 105^0C for 24 h. The coconut shells were ground using a Peppink hammer mill. The biosorbent was then sieved to obtain the 0.5–1.4 mm fraction used throughout this study. Table 7-1 shows the physical characteristics of the biosorbent. SEM-EDX, FTIR and point of zero charge (PZC) analysis of the coconut shell prior to and after Cu(II) sorption were presented in Acheampong et al. (2011) (Chapter 4). The SEM-EDX characterisation of coconut shell before and after sorption showed irregular surface texture of the biosorbent and the presence of K^+ and Mg^{2+} that were replaced by copper during sorption (Acheampong et al., 2011). The FTIR analysis before and after sorption suggested the involvement of amino, amide, carboxylic, hydroxyl and carbonyl groups in copper binding through the ion exchange mechanism. The point of zero charge measured (Table 7-1) showed that electrostatic attraction forces played a role in the copper removal.

Table 7-1: Physical characteristics of the coconut shell used in this study (Acheampong et al., 2011)

Parameter	Value
Specific volume* ($cm^3 g^{-1}$)	0.9
Porosity (%)	55
Density ($g\ cm^{-3}$)	1.35
Specific surface area ($m^2 g^{-1}$)	0.4

*Including the pore volume

7.2.2   Chemicals
All reagents used in this study were of analytical grade. A stock solution (1000 mg L^{-1}) of Cu(II) was prepared by dissolving the appropriate amount of copper chloride (CuCl$_2$) in distilled water. Solutions of lower concentrations were prepared by diluting the stock solution with double distilled water. Sodium hydroxide (NaOH) and hydrochloric acid (HCl) were used for pH adjustment.

7.2.3   Fixed-bed column set-up and experiments
The experimental set-up along with the fixed-bed column investigated in this study is shown in Figure 7-1. The column was constructed out of a Perspex cylinder (10 cm inner diameter and 80 cm long), packed with coconut shell (particle size 0.5 - 1.4 mm) to give a total effective bed depth of 20 cm. The biosorbent was supported in between a layer of glass beads at the top and a layer of gravel at the bottom. Four sampling ports were located 5 cm apart within the 20 cm packing height for drawing samples for Cu(II) analysis. For regulating the liquid flow inside the column, two controllable valves were used as shown in Figure 7-1.

The column performance of Cu(II) biosorption onto coconut shell was studied at different flow rates (10, 20 and 30 mL min^{-1}) and inlet Cu(II) concentrations (10, 20 and 50 mg L^{-1}) at an inlet pH of 7.0 ± 0.2. The effect of bed-depth was investigated at 5, 10, 15 and 20 cm, equivalent to 243, 486, 729 and 972 g of biosorbent, respectively. All continuous operations with the column were performed under down-flow mode. Each sample analysis was carried out in triplicate and results reported are average with ± 3% standard deviation error.

7.2.4   Analytical techniques
The concentration of Cu in the samples was measured using an atomic absorption spectrometer (AAS, Perkin Elmer, model AAnalist200), equipped with an air-acetylene flame. The detection limit of the AAS for Cu(II) is 0.01 mg L^{-1}. As a quality assurance procedure, a 3-point calibration curve through zero was applied with a blank solution as well as standard copper solutions of 0.5, 1.0 and 1.5 mg L^{-1}. A resulting linear calibration line ($R^2 \geq 0.996$) was the requirement for any sample measurement. Furthermore, after every 10 measurements, the 1.0 mg L^{-1} standard solution was checked for accuracy. The calibration was repeated in cases where incorrect values were obtained. Finally, the standard solution (1.3 mg L^{-1}) for the required absorbance (0.20) was checked as part of the quality assurance process.

The solution pH was measured with a SenTix21 pH electrode (WTW model pH323). The pH meter was calibrated using buffer solutions with pH values of 4.0, 7.0 and 10.0.

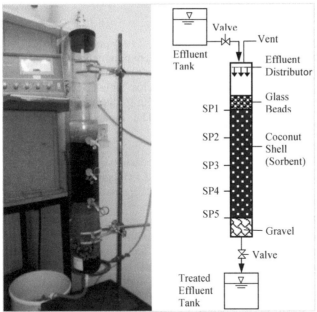

Figure 7-1: Photo and schematic representation of the laboratory-scale fixed-bed column setup. SP = sampling points

7.2.5   Calculations

The maximum column capacity, q_{total} (mg) for a given set of conditions in the column was calculated from the area under the plot of adsorbed Cu(II) concentration, C_{ad} (mgL^{-1}), versus time as given by the equation (Ahmad and Hameed, 2010):

$$q_{total} = \frac{QA}{1000} = \frac{Q}{1000} \int_{t=0}^{t=t_{total}} C_{ad}\, dt \tag{7-1}$$

where $C_{ad} = C_i$ - C_e (mg L^{-1}), t_{total} is the total flow time (min), Q is the flow rate (mL min^{-1}) and A is the area under the breakthrough curve (cm^2).

The equilibrium uptake ($q_{e(exp)}$), i.e. the amount of Cu(II) adsorbed (mg) per unit dry weight of adsorbent (mg g^{-1}) in the column, was calculated from Eq. 7-2 (Mohan and Sreelakshmi, 2008; Martin-Lara et al., 2012):

$$q_{eq(exp)} = \frac{q_{total}}{W} \tag{7-2}$$

where W is the total dry weight of coconut shell in the column (g).
The total volume treated, V_{eff} (mL), was calculated from Eq. 7-3 (Futalan et al., 2011):

$$V_{eff} = Q t_{total} \tag{7-3}$$

7.3   Dynamic models

For the successful design of a column adsorption process, it is important to predict the concentration-time profile or breakthrough curve for effluent parameters. A number of mathematical models have been developed for use in the design of continuous fixed bed sorption columns. In this work, the Bed Depth Service Time (BDST), Thomas, Yoon-Nelson and Clark models were used in predicting the behaviour of the breakthrough curve because of their effectiveness. The model's equations are presented in Table 7-2.

Among the various design approaches, the BDST approach based on the Bohart-Adams model (Bhart and Adams, 1920) is widely used (Mohan and Sreelakshmi, 2008; Luo et al., 2011a). It assumes that the rate of adsorption is governed by the surface reaction between the adsorbate and the unused capacity of the adsorbent. The BDST model describes the relation between the breakthrough time, often called the service time of the bed and the packed-bed depth of the column. The advantage of the BDST model is that any experimental test can be reliably scaled up to other flow rates and inlet solute concentrations without further experimental test (Vijayaraghavan and Prabu, 2006). The BDST model rate constant (K_a) is not significantly affected by the change in flow rate, and thus the intercept n of the BDST equation (Table 7-2) remains unchanged when flow rate is changed (Vijayaraghavan and Prabu, 2006). However, the slope of the equation does change with change in flow rate and hence given by:

$$\text{New slope} = \text{old slope}\left(\frac{Q_{old}}{Q_{new}}\right) \qquad (7\text{-}4)$$

In contrast, the change in inlet solute concentration (C_0) usually results in a change in slope and intercept of the BDST equation. The new slope and intercept values can be determined from (Vijayaraghavan and Prabu, 2006):

$$\text{New slope} = \text{old slope}\left(\frac{C_{0,old}}{C_{0,new}}\right) \qquad (7\text{-}5)$$

$$\text{New intercept} = \text{old intercept}\left(\frac{C_{0,old}}{C_{0,new}}\right)x\left(\frac{\ln\left[\left(C_{0,new}/C_B\right)\right]-1}{\ln\left[\left(C_{0,old}/C_B\right)\right]-1}\right) \qquad (7\text{-}6)$$

The Thomas model (Thomas, 1944) is one of the most widely used models in describing the column performance and prediction of breakthrough curves. The model follows the Langmuir kinetics of adsorption-desorption. It assumes negligible axial dispersion in the column adsorption since the rate driving force obeys the second-order reversible kinetics (Futalan et al., 2011).

Table 7-2: Summary of the models used to evaluate the breakthrough curves in this study

Model Name	Linearised model equations	Remarks	Reference
Bed Depth Service Time (BDST)	$$t = \frac{N_0}{C_0 F} Z - \frac{1}{K_a C_0} \ln\left(\frac{C_0}{C_B} - 1\right) \quad (7\text{-}7)$$	A plot of t versus bed depth, Z, should yield a straight line where N_0 and K_a can be evaluated from the slope and intercept, respectively	Bohart and Adams (1920); Luo et al. (2011a); Han et al. (2009)
	$$Z_0 = \frac{F}{K_a N_0} \ln\left(\frac{C_0}{C_B} - 1\right) \quad (7\text{-}8)$$	Z_0 is the critical bed depth	
Thomas	$$\ln\left(\frac{C_0}{C_t} - 1\right) = \frac{k_{Th} q_0 M}{Q} - k_{Th} C_0 t \quad (7\text{-}9)$$	A plot of $\ln[C_0/C_t - 1]$ against t gives a straight line from which the values of k_{Th} and q_0 are determined from the intercept and the slope, respectively	Thomas (1944)
Yoon-Nelson	$$\ln\left(\frac{C_t}{C_0 - C_t}\right) = k_{YN} t - \tau k_{YN} \quad (7\text{-}10)$$	A plot of $\ln[C_t/(C_0 - C_t)]$ against time t gives a straight line from which the values of k_{YN} and τ can be determined from the slope and the intercept, respectively	Yoon and Nelson (1984)
Clark	$$\ln\left[\left(\frac{C_0}{C}\right)^{n-1} - 1\right] = \ln A - rt \quad (7\text{-}11)$$	A and r are determined from the slope and the intercept of plots of $\ln((C_0/C)^{n-1} - 1))$ versus t	Clark (1987); Pakshirajan and Swaminathan (2006)

Yoon and Nelson (Yoon and Nelson, 1984) developed a model to investigate the breakthrough behaviour of adsorbate gases on activated carbon. The model was based on the assumption that the rate of decrease in the probability of adsorption of each adsorbate molecule is proportional to the probability of the adsorbate adsorption and the adsorbate breakthrough on the adsorbent (Chen et al., 2012). The kinetic equation developed by Clark (Clark, 1987) assumes that the sorption behaviour of pollutants follows the Freundlich adsorption isotherm, and the sorption rate is determined by the external mass transfer step. The model is used for the simulation of breakthrough curves.

7.4   Results

7.4.1   Effect of flow rate on Cu(II) biosorption

The breakthrough curves obtained at different flow rates, constant inlet Cu(II) concentration and bed-depth are presented in Figure 7-2. The curves show that the breakthrough time as well as the exhaustion time increased with decrease in flow rate. The slope of the plots from breakthrough time to exhaustion time increased as the flow rate was increased from 10 to 30 mL min^{-1}, showing that the breakthrough curve becomes steeper as the flow rate increased. A lower flow rate results in a higher residence time in the column and vice versa (Figure 7-2). An increase in the flow rate reduced the volume of effluent treated before the bed became saturated and therefore decreased the service time of the bed (Figure 7-2). A decrease in the contact time between the Cu(II) ions and the coconut shells at higher linear flow rates accounted for this observation. Figure 7-2 shows that breakthrough occurs faster at a higher flow rate.

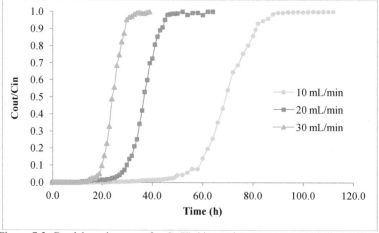

Figure 7-2: Breakthrough curves for Cu(II) biosorption onto coconut shell at different flow rates. Experimental condition: bed depth = 20 cm, inlet Cu(II) concentration = 10 mg L^{-1}, particle size = 0.5 - 1.4 mm, temperature = 30 ± 1 ^{0}C, influent pH =7.0 ± 0.2.

7.4.2   Effect of bed depth on Cu(II) biosorption

Figure 7-3 shows the breakthrough curves of Cu(II) biosorption onto coconut shell obtained at different bed depths with a Cu(II) inlet concentration of 10 mg L^{-1} and a constant flow rate of 10 mL min^{-1}. Four bed depths (5, 10, 15 and 20 cm), corresponding to 243, 486, 729 and 972 g dry weight of coconut shell, respectively, were investigated. The breakthrough curves (Figure 7-3) show that the breakthrough time and exhaustion time increased with increase in bed depth from 5 to 20 cm.

7.4.3   Effect of inlet Cu(II) concentration on Cu(II) biosorption

The effect of varying the inlet Cu(II) concentration from 10 to 50 mg L^{-1} on the shape of the breakthrough curves was studied at a constant biosorbent bed depth (20 cm) and feed flow rate (10 mL min^{-1}). The resulting breakthrough curves are presented in Figure 7-4.

The curves show that the breakthrough time decreased with increasing inlet Cu(II) concentration. The larger the C_0, the steeper the breakthrough curve and the shorter the breakthrough time (Figure 7-4). An increase in the inlet concentration reduced the treated volume before the bed gets saturated, since a high Cu(II) concentration may saturate the adsorbent more quickly, thereby decreasing the operation time. Decreasing the Cu(II) concentration increases the volume of the feed Cu(II) solution that can be treated, shifting the breakthrough curve to the right (Figure 7-4). During the first 30 h of the column operation, the value of C_{out}/C_{in} reached 1.0, 0.55 and 0.01 when the inlet concentrations were 50 mg L^{-1}, 20 mg L^{-1} and 10 mg L^{-1}, respectively.

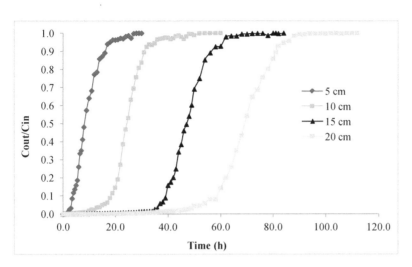

Figure 7-3: Breakthrough curves for Cu(II) biosorption onto coconut shell at different bed heights. Experimental conditions: inlet Cu(II) concentration = 10 mg L^{-1}, flow rate = 10 mL min^{-1}, particle size = 0.5 - 1.4 mm, temperature = 30 ± 1 ^{0}C, feed pH =7.0 ± 0.2.

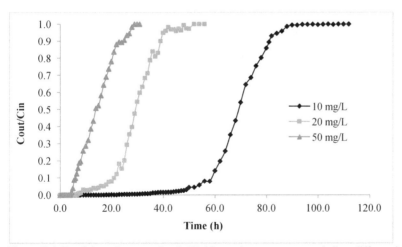

Figure 7-4: Breakthrough curves for Cu(II) biosorption onto coconut shell at different initial Cu(II) concentrations. Experimental conditions: flow rate = 10 mL min^{-1}, bed depth = 20 cm, particle size = 0.5 - 1.4 mm, temperature = 30 ± 1 ^{0}C, influent pH =7.0 ± 0.2.

7.4.4   Cu(II) uptake at different column operation parameters

Data collected during the laboratory tests serve as the basis for the design of a full scale biosorption column. The column data obtained during the experimental run are presented in Table 7-3. As the flow rate increases, the volume of influent treated increased while the uptake decreased (Table 7-3). The best column operation conditions for Cu(II) biosorption by coconut shell were 10 mg L^{-1} Cu(II) inlet concentration, 20 cm bed depth and 10 mL min^{-1} flow rate, which yielded a service time to breakthrough of 60 h (Figure 7-4) for meeting the Ghana EPA limit of 1 mg L^{-1} Cu(II) concentration in industrial wastewater for discharge.

Table 7-3: Uptake of Cu(II) at different flow rates. Experimental conditions: particle size = 0.5 - 1.4 mm, bed depth = 20 cm, initial Cu(II) concentration = 10 mg L^{-1}, temperature = 30 ± 1 ^{0}C, inlet pH = 7.0 ± 0.2.

Z (cm)	Q (mL min^{-1})	C_0 (mg L^{-1})	V_{eff} (mL)	q_{total} (mg)	$q_{e(exp)}$ (mg g^{-1})	Z_m (cm)
20	10	10	67200	6720	7.25	8.93
20	20	10	62400	6240	6.73	7.31
20	30	10	57600	5760	6.21	6.89

7.4.5   Dynamic models

7.4.5.1   Bed Depth Service Time (BDST)

Fig. 7-5 shows the plot of the service time versus bed height for Cu(II) biosorption onto coconut shell for 10% and 50% saturation. The BDST parameters calculated from the slopes and the intercepts of the plots are presented in Table 7-3. The variation of the service time with bed depth is linear for the 10% and 50% plots (Figure 7-5), with very high correlation coefficient (R^2) values, indicating the validity of the BDST model for this biosorption system studied.

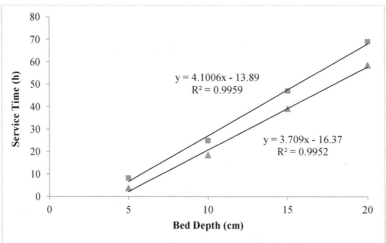

Figure 7-5: BDST model plots for Cu(II) biosorption onto coconut shell at 10% saturation (■) and 50% saturation (▲); at different bed heights. Experimental conditions: flow rate = 10 mL min^{-1}, initial Cu(II) concentration = 10 mg L^{-1}, particle size = 0.5 - 1.4 mm, Temperature = 30 ± 1 ^{0}C, inlet pH = 7.0 ± 0.2.

Table 7-4: BDST model parameters for the sorption of Cu(II) onto coconut shell at varying bed heights for 10% and 50% saturation. Experimental conditions: flow rate = 10 mL min^{-1}, initial Cu(II) concentration = 10 mg L^{-1}, particle size = 0.5 - 1.4 mm, temperature = 30 ± 1 ^{0}C, inlet pH = 7.0 ± 0.2.

BDST parameters	10% Saturation	50% Saturation
N$_0$ (mg L^{-1})	2,225	2,460
K$_a$ (L mg^{-1} h)	0.0134	0.0049
Z$_0$ (cm)	12.51	10.61
R^2	0.996	0.995

7.4.5.2   The Thomas, Yoon-Nelson and Clark models

The Thomas model was applied to the experimental data with respect to the initial Cu(II) concentration and flow rate. The model parameters were determined from the linear plot (data not shown) of Eq. 7-9 and are presented in Table 7-5. Table 7-5 shows that k$_{TH}$ increased with increasing flow rate and initial Cu(II) concentration. Similarly, q$_0$ increased with flow rate and the initial Cu(II) concentration over the entire sorption period (Table 7-5). A good fit (R^2 > 0.96) was obtained for all the plots.

The values of the Yoon-Nelson parameters (k$_{YN}$ and τ) were determined from the plot of ln[(C$_t$/(C$_0$-C$_t$)] versus t at various operating conditions (Table 5). The k$_{YN}$ values increased with increasing flow rates, while the τ values (time for 50% breakthrough) decreased as the flow rate increased. A similar trend was observed when the initial Cu(II) concentration was increased (Table 7-5). The R^2 values (Table 7-5) indicate a good fit in all cases, showing that the Yoon-Nelson model can be used to describe the Cu(II)-coconut shell biosorption system.

The values of the Clark parameters A and r are presented in Table 7-5. The applicability of the Clark model to this system is very feasible in the examined range of flow rate, bed depth and initial Cu(II) concentration for which their dependence on parameters *A* and *r* have been established (Table 7-5). As expected, the values of r increase with increasing bed height, flow rate and initial Cu(II) concentration, while the A values are inversely related to the experimental conditions. The testing of experimental breakthrough curves by the linearised Clark model gave excellent fitting of the experimental data as evident from the high R^2 values (Table 7-5).

Table 7-5: The Thomas, Yoon-Nelson and Clark models parameters for Cu(II) biosorption onto coconut shell at different bed heights, flow rate and inlet concentration. Experimental conditions: particle size = 0.5 - 1.4 mm, temperature = 30 ± 1 ^{0}C, inlet pH = 7.0 ± 0.2.

Experimental conditions			Thomas model Parameters			Yoon-Nelson model parameters			Clark parameters		model
Z (cm)	Q (mL min^{-1})	C$_0$ (mg L^{-1})	q$_0$ (mg g^{-1})	k$_{Th}$ (mL mg^{-1} min^{-1})	R^2	k$_{YN}$ (mg g^{-1})	τ (min^{-1})	R^2	R (min^{-1})	A (-)	R^2
15	10	10	43.022	0.0003	0.961	0.003	2,739	0.961	0.0031	2.318	0.999
15	20	10	20.979	0.0005	0.984	0.005	1,480	0.985	0.0057	2.340	0.998
15	30	10	13.186	0.0008	0.993	0.008	925	0.993	0.0088	2.179	0.997
20	10	10	53.498	0.0002	0.967	0.004	3,936	0.982	0.0037	2.359	0.998
20	10	20	22.438	0.0003	0.964	0.004	1,738	0.962	0.0038	1.902	0.999
20	10	50	10.365	0.0004	0.972	0.005	877	0.974	0.0044	1.482	0.999

7.5   Discussion

7.5.1   Cu(II) uptake in the column

This study showed that the sorption uptake capacity of the column (7.25 mg g^{-1}) was 14.5 times better than that obtained in a batch system (Acheampong et al., 2011). The increased capacity of the column method can be attributed to the continuously increasing concentration gradient in the interface of the adsorption zone as it passes through the column, whereas the gradient concentration decreases with time in batch systems (Sousa et al., 2010; Martin-Lara et al., 2012). A characterisation study on the coconut shell prior to and after biosorption (Chapter 4) showed that hydroxyl and carboxylic functional groups were involved in the removal of Cu(II) from aqueous solutions by this biosorbent, besides micro precipitation and electrostatic attraction forces (Acheampong et al., 2011). The results obtained by Sousa et al. (2010) for Ni(II), Cd(II), Zn(II) and Pb(II) ions using NaOH treated green coconut shell suggest that a lower pH of 5 is required for optimal removal of the studied metals as compared to the pH of 7 ± 0.2 used in this study. They reported a reduction in removal efficiency during cyclical application, possibly due to the effect of the chemical treatment.

7.5.2   Effect of operating parameters on breakthrough curves

The column performed well at the lowest flow rate because at the lower flow rate the residence time of the feed solution was increased, allowing Cu(II) ions to diffuse into the pores of the coconut shell by means of intra-particle diffusion (Luo et al., 2011a). A high flow rate means insufficient time for Cu(II) ions to diffuse into the pores of the coconut shell, leading to a low uptake capacity and removal efficiency (Ko et al., 2000; Vijayaraghavan et al., 2004). This may be due to the solute leaving the column before the equilibrium could be achieved (Singh and Pant, 2006).

A higher bed height allowed for a higher Cu(II) uptake (Table 7-3). As the bed height increased, the Cu(II) ions had more time to contact with more coconut shell particles, resulting in a higher uptake of Cu(II) in the column. Hence, when the bed height increases, the maximum adsorption capacity of the column also increases (Malkoc and Nuhoglu, 2006). According to Singh et al. (2009) and Han et al. (2006a), the increase in the number of binding sites resulting from the increase in the adsorption surface area accounted for the increased uptake of Cu(II). The slight increase in the slope of the breakthrough curves (Figure 7-3) with increasing bed height resulted in a broadened mass transfer zone. When the bed depth is reduced, axial dispersion phenomena predominate in the mass transfer and reduce the diffusion of metallic ions (Taty-Costodes et al., 2005). At low bed depth in the column, the Cu(II) ions did not have enough time to diffuse into the coconut shell bed. Consequently, a reduction in the total treatment time, and hence volume of Cu(II) solution treated, was observed (Figure 7-3).

The driving force for adsorption is the concentration gradient between the sorbates on the adsorbent and the solution (Oguz and Ersoy, 2010). A high concentration, therefore, provides a large driving force for the adsorption process. However, the saturation of the adsorbent requires much more time. But due to the high driving force, breakthrough is reached before all the active sites of the coconut shell could be occupied by Cu(II) ions, leading to a shorter breakthrough time (Figure 7-4). The diffusion process is concentration dependent because a change of concentration gradient affects the saturation rate and the breakthrough time (Mondal, 2009). The dependence of the diffusion process on the sorbate concentration was reported for the sorption of Cu(II) and Cd(II) ions onto bone char (Ko et al., 2000), Cu(II) onto expanding rice husk (Luo et al., 2011a) and Cu(II) onto carboxylic acid functionlized deacetylated glucomannan (Luo et al., 2011b).

7.5.3   Design of fixed bed sorption columns

The most important criterion in the design of fixed-bed adsorption systems is the prediction of the column breakthrough or the shape of the adsorption wave front, which determines the operating life-span of the bed. The BDST model in general gave an idea of the efficiency of the column under constant operating conditions for achieving a desired Cu(II) breakthrough level in the column. The values of sorption rate constant (K_a) and Cu(II) sorption capacity (N_0) of the column (Table 7-4), estimated using this model, are high and match closely for both 10 and 50% Cu(II) breakthrough in the column. In addition, these parameter values are also in good agreement with our previous study on batch kinetics of Cu(II) biosorption using coconut shells (Acheampong et al., 2011), thus indicating a favourable removal of the metal even under the different continuous column operating conditions adopted in this study. The BDST model parameters are useful in scaling up the process for other flow rates without further experimental runs (Vijayaraghavan and Prabu, 2006). According to Singh et al. (2009), a 50% breakthrough curve between t and Z must pass through the origin, which was not the case in this study. This showed that the adsorption of Cu(II) onto coconut shell is governed by a more complex mechanism and the sorption process is limited by the intraparticle diffusion step, with the film diffusion contributing significantly in the initial stage of sorption (Singh et al., 2009). The rate constant K_a, which is a measure of the rate of transfer of solute from the fluid phase to the solid phase (Chandra-Sekhar et al., 2003; Al-Degs et al., 2009), largely influenced the breakthrough phenomenon in the column study. For smaller values of K_a, a relatively longer bed is required to avoid breakthrough whereas the breakthrough can be eliminated even in smaller bed heights when the value of K_a is high (Chandra-Sekhar et al., 2003).

The process parameter that is likely to change with time is the bed capacity, N_0. This parameter is used to predict the performance of the bed if there is a change in the initial solute concentration, C_0, to a new value. At a constant flow velocity of the feed, a greater N_0 and a smaller Z_0 indicate a higher efficiency for an adsorption material, since K_a and N_0 are inversely proportional to Z, and the product of K_a and N_0 is a unique constant for a given adsorbent (Luo et al., 2011a). From Table 7-4, N_0 is much greater than Z_0 for both the 10% and 50% saturation cases, indicating that coconut shell is highly efficient in removing Cu(II) from aqueous environments. Once the BDST column design parameters are found, fixed bed columns can be designed for deferent flow rates and initial solute concentrations without further experimental runs (Vijayaraghavan and Prabu, 2006).

The adsorption capacity predicted by the Thomas model (Table 7-5) did not agree well with those obtained from the experimental results (Table 7-3). This showed that the Thomas model can not sufficiently describe the biosorption system in this study and is therefore not a suitable model for the design of the sorption column, although the R^2 value indicates a good fit. The thickness of the liquid film on the adsorbent surface has a direct effect on the mass transfer resistance (Futalan et al., 2011). The higher flow rates enhance the mass transfer of the Cu(II) ions from the liquid film to the coconut shell surface, resulting in earlier saturation of the adsorbent bed. The decrease in q_0 as the flow rate was increased (Table 7-5) was due to the insufficient time for the Cu(II) ions to diffuse into the coconut shell bed. The driving force for adsorption is the concentration difference between the Cu(II) ions on the coconut shell and the Cu(II) remaining in the solution (Luo et al., 2011a). Therefore, a decrease in initial Cu(II) concentration lead to an increase in the value of q_0 and a decrease in k_{TH}. Designing sorption columns with the Thomas model should utilise low flow rates as well as inlet Cu(II) concentrations for optimal Cu(II) uptake. Nonetheless, the disagreement between the sorption capacities calculated experimentally and that predicted by the model makes it inappropriate for the coconut shell columns studied.

The Yoon-Nelson rate constant k_{YN} follows the same trend as the Thomas model rate constant k_{TH} (Table 7-5). The time required for 50% of sorbate breakthrough (τ) obtained from the Yoon-Nelson model agreed well with the experimental data at all conditions examined. Consequently, the Yoon-Nelson model gave a good representation of the Cu(II)-coconut shell system. The Yoon-Nelson model has been used successfully to predict the time required for 50% sorbate breakthrough of the biosorption of Cu(II) ions onto *Sagassum wightii* biomass (Vijayaraghavan and Prabu, 2006), sorption of Cu(II) by rice husk based activated carbon (Yahaya et al., 2011), removal of furfural using activated carbon (Singh et al., 2009), biosorption of reactive black 5 (Vijayaraghavan and Yun, 2008), sorption of Cr(VI) by modified corn stalk (Chen et al., 2011), removal of Cr(VI) by thermally activated weed *Salvinia cucullata* (Baral et al., 2009), transport and fate of Pb(II), Cd(II), Cr(VI) and As(V) in soil (Zhao et al., 2009), and biosorption of La(III) and Na(III) by *Sargassum sp.* (Oliveira et al., 2012). These results confirmed that the model set of equations can be used as a suitable mathematical representation of the biosorption process carried out in continuous flow fixed bed columns.

Among the various breakthrough models tested in this study, the Clark model could be considered a more refined model as it involves both the mass transfer and equilibrium adsorption in predicting the breakthrough phenomena. Our earlier work dealing with a batch study on Cu(II) biosorption using coconut shell showed that the Freundlich model provided a good fit to the experimental equilibrium data (Acheampong et al., 2011), and, therefore, the previously obtained Freundlich constant 'n' value was applied to estimate the Clark model parameters in this study. It is evident that the experimental breakthrough curve of Cu(II) at the different column operating conditions was well predicted by the Clark model over the entire time period (Table 7-5). Thus, the entire breakthrough curve can be designed accurately with the Clark model for continuous removal of Cu(II) by coconut shell in fixed bed columns. This study showed that for optimal column operation, the design parameters A and r should have values of 2.359 and 3.7×10^{-3} min^{-1}, respectively. The estimated values of the parameter 'r' were low and close to each other indicating a quick and effective Cu(II) mass transfer in the dynamically operated column for its removal by biosorption. These results are in agreement with those of Sag and Aktay (2001) who reported a similar order of magnitude for the parameter 'r' in chromium (VI) sorption by chitin in a packed column. Ghribi and Chlendi (2011) reported A and r values of 2.494 and 7.9×10^{-3}, respectively, for the sorption of organic dyes onto natural clay. The linearised Clark equation has been successfully applied for paint removal using calcium chloride treated beech sawdust (Batzias and Sidiras, 2004), sorption of Pb(II) ions by natural zeolite (Medvidović et al., 2008) and sorption of furfural by activated carbon (Singh et al., 2009), suggesting that the equation is a good mathematical tool for biosorption column design. The linear regression analysis method is faster, easier and less complicated (Medvidović et al., 2008). The non-linear Clark equation has also been successfully applied in modelling the biosorption of phenol on active mud (Aksu and Gönen, 2004), biosorption of organic pollutants by biomass (Aksu, 2005), removal of Cr(VI) by Chitin (Sag and Aktay, 2001), biosorption of Cu(II), Pb(II) and Cd(II) using *Phanerochaete chrysosporium* (Pakshirajan and Swaminathan, 2006). This means that the Clark model is useful in designing sorption columns for a variety of sorbent-sorbate systems for continuous operation. Medvidović et al. (2008) applied the linearised and the non-linear equations to the sorption of Pb(II) ions by natural zeolite and concluded that both equations yielded approximately the same values of A and r, which meant that they are acceptable for column design.

7.5.4   Stability and reusability of the biosorbent

Biosorbent stability and reusability are significant aspects in column operations. In a previous sorption-desorption study (Chapter 5), no physical deterioration, loss of dry mass or mechanical strength of the coconut shell was observed after eight sorption-desorption cycles using 0.05 M HCl as eluant (Acheampong et al., 2013). This observed mechanical stability and stiffness of the coconut shell make it suitable for fixed bed column applications. A desorption efficiency of 98% was reported with no loss of uptake capacity of the coconut shell after multiple reuse (Acheampong et al., 2013). The fast desorption process recorded (60 min) coupled with the good mechanical strength makes coconut shell suitable for heavy metal removal from wastewater in fixed bed columns. Sousa et al. (2010), however, observed a reduction in the removal capacity of the NaOH treated green coconut shell after five cycles of sorption and desorption. They reported a decrease in efficiency of 50%, 70%, 67% and 76% for Pb(II), Ni(II), Cd(II) and Zn(II), respectively, indicating that the chemical treatment had compromised the reusability potential of the sorbent. Further research is required to investigate if modification of the coconut shell material, either chemically or physically, can considerably improve the Cu(II) uptake capacity of the coconut shell without compromising its regeneration and reusability potential. Also the optimisation of the cost effective recovery of the copper from the concentrated solution obtained during the regeneration of the loaded sorbent, e.g. through electrochemical techniques such as electro-winning, needs to be investigated.

7.6   Conclusions

Fixed-bed biosorption systems were found to perform better for Cu(II) uptake by coconut shell at a lower Cu(II) inlet concentration, lower feed flow rate and higher coconut shell bed-depth. Under these optimum conditions, the service time to breakthrough and Cu(II) concentration were 58 h and 0.8 mg L^{-1}, respectively, after which the Cu(II) concentration in the effluent exceeded the 1 mg L^{-1} discharge limit set by the Ghana Environmental Protection Agency (EPA). Among the various models applied to describe the metal breakthrough in the column, the BDST model was able to describe the data only up to 10 % and 50 % saturation, whereas prediction of the Cu(II) uptake capacity by the Thomas model was poor. The Yoon-Nelson model predicted the time required for 50% of sorbate breakthrough well and can be applied to the entire breakthrough curve. However, the entire breakthrough curve was best predicted by the Clark model. The design of a continuous fixed bed column treatment system for copper laden wastewater can thus be achieved using the BDST, Yoon-Nelson and Clark breakthrough models.

7.7   Acknowledgements

The authors acknowledge funding from the Netherlands Government under the Netherlands Fellowship Programme (NFP), NUFFIC award (2009-2013), Project Number: 32022513. We also acknowledge funding from the Staff Development and Postgraduate Scholarship Scheme (Kumasi Polytechnic, Ghana) and the UNESCO-IHE Partner Research Fund, UPaRF III research project PRBRAMD (No. 101014).

7.8   References

Acheampong, M.A., Dapcic, A.D., Yeh, D. and Lens, P.N.L., 2013. Cyclic sorption and desorption of Cu(II) onto coconut shell and iron oxide coated sand. Separation Science and Technology (DOI:10.1080/01496395.2013.809362) (In press).
Acheampong, M.A., Pereira, J.P.C., Meulepas, R.J.W. and Lens, P.N.L., 2011. Biosorption of Cu(II) onto agricultural materials from tropical regions. Journal of Chemical Technology & Biotechnology, 86(9): 1184-1194.
Ahmad, A.A. and Hameed, B.H., 2010. Fixed-bed adsorption of reactive azo dye onto granular activated carbon prepared from waste. Journal of Hazardous Materials, 175(1-3): 298-303.
Aksu, Z. and Gönen, F., 2004. Biosorption of phenol by immobilized activated sludge in a continuous packed bed: prediction of breakthrough curves. Process Biochemistry, 39(5): 599-613.
Aksu, Z., 2005. Application of biosorption for the removal of organic pollutants: a review. Process Biochemistry, 40(3–4): 997-1026.
Al-Degs, Y.S., Khraisheh, M.A.M., Allen, S.J. and Ahmad, M.N., 2009. Adsorption characteristics of reactive dyes in columns of activated carbon. Journal of Hazardous Materials, 165(1-3): 944-949.
Amarasinghe, B.M.W.P.K. and Williams, R.A., 2007. Tea waste as a low cost adsorbent for the removal of Cu and Pb from wastewater. Chemical Engineering Journal, 132(1-3): 299-309.
Apiratikul, R. and Pavasant, P., 2008. Batch and column studies of biosorption of heavy metals by Caulerpa lentillifera. Bioresource Technology, 99(8): 2766-2777.
Baral, S.S. et al., 2009. Removal of Cr(VI) by thermally activated weed Salvinia cucullata in a fixed-bed column. Journal of Hazardous Materials, 161(2–3): 1427-1435.
Batzias, F.A. and Sidiras, D.K., 2004. Dye adsorption by calcium chloride treated beech sawdust in batch and fixed-bed systems. Journal of Hazardous Materials, 114(1–3): 167-174.
Bohart, G.S. and Adams, E.Q., 1920. Some aspects of the behavior of charcoal with respect to chlorine.1. Journal of the American Chemical Society, 42(3): 523-544.
Chandra Sekhar, K., Kamala, C.T., Chary, N.S. and Anjaneyulu, Y., 2003. Removal of heavy metals using a plant biomass with reference to environmental control. International Journal of Mineral Processing, 68(1-4): 37-45.
Chen, C.-Y., Yang, C.-Y. and Chen, A.-H., 2011. Biosorption of Cu(II), Zn(II), Ni(II) and Pb(II) ions by cross-linked metal-imprinted chitosans with epichlorohydrin. Journal of Environmental Management, 92(3): 796-802.
Chen, S. et al., 2012. Adsorption of hexavalent chromium from aqueous solution by modified corn stalk: A fixed-bed column study. Bioresource Technology, 113(0): 114-120.
Clark, R.M., 1987. Evaluating the cost and performance of field-scale granular activated carbon systems. Environmental Science & Technology, 21(6): 573-580.
Fagundes-Klen, M. et al., 2010. Copper Biosorption by Biomass of Marine Alga: Study of Equilibrium and Kinetics in Batch System and Adsorption/Desorption Cycles in Fixed Bed Column. Water, Air, & Soil Pollution, 213(1): 15-26.
Futalan, C.M., Kan, C.-C., Dalida, M.L., Pascua, C. and Wan, M.-W., 2011. Fixed-bed column studies on the removal of copper using chitosan immobilized on bentonite. Carbohydrate Polymers, 83(2): 697-704.
Ghana EPA, 2010. Environmental performance rating and disclosure: report on the performance of mining and manufacturing companies. Environmental Protection Agency, Accra, Ghana.
Ghribi, A. and Chlendi, M., 2011. Modeling of fixed bed adsorption: Application to the adsorption of

an organic dye, Asian Journal of Textile, 1(4): 161-171.

Han, R. et al., 2006a. Biosorption of copper(II) and lead(II) from aqueous solution by chaff in a fixed-bed column. Journal of Hazardous Materials, 133(1-3): 262-268.

Han, R. et al., 2009. Characterization and properties of iron oxide-coated zeolite as adsorbent for removal of copper(II) from solution in fixed bed column. Chemical Engineering Journal, 149(1-3): 123-131.

Ko, D.C.K., Porter, J.F. and McKay, G., 2000. Optimised correlations for the fixed-bed adsorption of metal ions on bone char. Chemical Engineering Science, 55(23): 5819-5829.

Lu, X., Wang, L., Lei, K., Huang, J. and Zhai, Y., 2009. Contamination assessment of copper, lead, zinc, manganese and nickel in street dust of Baoji, NW China. Journal of Hazardous Materials, 161(2-3): 1058-1062.

Luo, X., Deng, Z., Lin, X. and Zhang, C., 2011a. Fixed-bed column study for Cu^{2+} removal from solution using expanding rice husk. Journal of Hazardous Materials, 187(1-3): 182-189.

Luo, X., Liu, F., Deng, Z. and Lin, X., 2011. Removal of copper(II) from aqueous solution in fixed-bed column by carboxylic acid functionalized deacetylated konjac glucomannan. Carbohydrate Polymers, 86(2): 753-759.

Malkoc, E. and Nuhoglu, Y., 2006. Fixed bed studies for the sorption of chromium(VI) onto tea factory waste. Chemical Engineering Science, 61(13): 4363-4372.

Martín-Lara, M.A., Blázquez, G., Ronda, A., Rodríguez, I.L. and Calero, M., 2012. Multiple biosorption–desorption cycles in a fixed-bed column for Pb(II) removal by acid-treated olive stone. Journal of Industrial and Engineering Chemistry, 18(3): 1006-1012.

Medvidović, N.V., Perić, J. and Trgo, M., 2008. Testing of Breakthrough Curves for Removal of Lead Ions from Aqueous Solutions by Natural Zeolite-Clinoptilolite According to the Clark Kinetic Equation. Separation Science and Technology, 43(4): 944-959.

Mohan, S. and Sreelakshmi, G., 2008. Fixed bed column study for heavy metal removal using phosphate treated rice husk. Journal of Hazardous Materials, 153(1-2): 75-82.

Mondal, M.K., 2009. Removal of Pb(II) ions from aqueous solution using activated tea waste: Adsorption on a fixed-bed column. Journal of Environmental Management, 90(11): 3266-3271.

Naja, G. and Volesky, B., 2006. Multi-metal biosorption in a fixed-bed flow-through column. Colloids and Surfaces A: Physicochemical and Engineering Aspects, 281(1-3): 194-201.

Nayek, S., Gupta, S. and Saha, R.N., 2010. Metal accumulation and its effects in relation to biochemical response of vegetables irrigated with metal contaminated water and wastewater. Journal of Hazardous Materials, 178(1-3): 588-595.

Oguz, E. and Ersoy, M., 2010. Removal of Cu^{2+} from aqueous solution by adsorption in a fixed bed column and Neural Network Modelling. Chemical Engineering Journal, 164(1): 56-62.

Oliveira, R.C., Guibal, E. and Garcia, O., 2012. Biosorption and desorption of lanthanum(III) and neodymium(III) in fixed-bed columns with Sargassum sp.: Perspectives for separation of rare earth metals. Biotechnology Progress, 28(3): 715-722.

Pakshirajan, K. and Swaminathan, T., 2006. Continuous Biosorption of Pb, Cu, and Cd by Phanerochaete chrysosporium in a Packed Column Reactor. Soil and Sediment Contamination: An International Journal, 15(2): 187-197.

Qaiser, S., Saleemi, A.R. and Umar, M., 2009. Biosorption of lead from aqueous solution by Ficus religiosa leaves: Batch and column study. Journal of Hazardous Materials, 166(2-3): 998-1005.

Sağ, Y. and Aktay, Y., 2001. Application of equilibrium and mass transfer models to dynamic removal of Cr(VI) ions by Chitin in packed column reactor. Process Biochemistry, 36(12): 1187-1197.

Shi, W.-y., Shao, H.-b., Li, H., Shao, M.-a. and Du, S., 2009. Progress in the remediation of hazardous heavy metal-polluted soils by natural zeolite. Journal of Hazardous Materials, 170(1): 1-6.

Singh, S., Srivastava, V.C. and Mall, I.D., 2009. Fixed-bed study for adsorptive removal of furfural by activated carbon. Colloids and Surfaces A: Physicochemical and Engineering Aspects, 332(1): 50-56.

Singh, T.S. and Pant, K.K., 2006. Experimental and modelling studies on fixed bed adsorption of As(III) ions from aqueous solution. Separation and Purification Technology, 48(3): 288-296.

Sousa, F.W. et al., 2010. Green coconut shells applied as adsorbent for removal of toxic metal ions using fixed-bed column technology. Journal of Environmental Management, 91(8): 1634-1640.

Stylianou, M.A., Inglezakis, V.J., Moustakas, K.G., Malamis, S.P. and Loizidou, M.D., 2007. Removal of Cu(II) in fixed bed and batch reactors using natural zeolite and exfoliated vermiculite as adsorbents. Desalination, 215(1-3): 133-142.

Taty-Costodes, V.C., Fauduet, H., Porte, C. and Ho, Y.-S., 2005. Removal of lead (II) ions from synthetic and real effluents using immobilized Pinus sylvestris sawdust: Adsorption on a fixed-bed column. Journal of Hazardous Materials, 123(1-3): 135-144.

Thomas, H.C., 1944. Heterogeneous Ion Exchange in a Flowing System. Journal of the American Chemical Society, 66(10): 1664-1666.

Tung, C.-C., Yang, Y.-M., Chang, C.-H. and Maa, J.-R., 2002. Removal of copper ions and dissolved phenol from water using micellar-enhanced ultrafiltration with mixed surfactants. Waste Management, 22(7): 695-701.

Vijayaraghavan, K. and Prabu, D., 2006. Potential of Sargassum wightii biomass for copper(II) removal from aqueous solutions: Application of different mathematical models to batch and continuous biosorption data. Journal of Hazardous Materials, 137(1): 558-564.

Vijayaraghavan, K. and Yun, Y.-S., 2008. Polysulfone-immobilized Corynebacterium glutamicum: A biosorbent for Reactive black 5 from aqueous solution in an up-flow packed column. Chemical Engineering Journal, 145(1): 44-49.

Vijayaraghavan, K., Jegan, J., Palanivelu, K. and Velan, M., 2004. Removal of nickel(II) ions from aqueous solution using crab shell particles in a packed bed up-flow column. Journal of Hazardous Materials, 113(1–3): 223-230.

Vilar, V.J.P., Loureiro, J.M., Botelho, C.M.S. and Boaventura, R.A.R., 2008. Continuous biosorption of Pb/Cu and Pb/Cd in fixed-bed column using algae Gelidium and granulated agar extraction algal waste. Journal of Hazardous Materials, 154(1–3): 1173-1182.

Wu, G. et al., 2010. A critical review on the bio-removal of hazardous heavy metals from contaminated soils: Issues, progress, eco-environmental concerns and opportunities. Journal of Hazardous Materials, 174(1-3): 1-8.

Yahaya, N.K.E.M., Abustana, I., Latiff, M.F.I.P.M., Bello, O.S. and Ahmad, M.A., 2011. Fixed-bed column study for Cu (II) removal from aqueous solutions using rice husk based activated carbon. International Journal of Engineering & Technology, 11(1): 248-252.

Yoon, Y.H. and Nelson, J.H., 1984. Application of Gas Adsorption Kinetics I. A Theoretical Model for Respirator Cartridge Service Life. American Industrial Hygiene Association Journal, 45(8): 509-516.

Zhao, X., Dong, D., Hua, X. and Dong, S., 2009. Investigation of the transport and fate of Pb, Cd, Cr(VI) and As(V) in soil zones derived from moderately contaminated farmland in Northeast, China. Journal of Hazardous Materials, 170(2–3): 570-577.

Chapter 8

8 Treatment of gold mining effluent in pilot fixed bed sorption systems

This was published as:
Acheampong, M.A., Lens, P.N.L., 2013. Treatment of gold mining effluent in pilot fixed bed sorption systems. Hydrometallurgy (Accepted)

Abstract

This paper studied the removal of heavy metals from gold mining effluent (GME) of the AngloGold Ashanti mine (Obuasi, Ghana) in continuous down-flow fixed bed columns using coconut shell and iron oxide-coated sand as sorbents operated at a temperature of 28 ± 2 ^{0}C and a constant flow-rate of 150 mL min^{-1}. The two-stage treatment system targeted the removal of copper and arsenic, but other heavy metals (iron, lead and zinc) present in the GME in very low concentrations were also removed, with removal efficiencies exceeding 98% for all metals in all the cases studied. A total of 14.8 m^3 of GME was treated in 1,608 h before arsenic breakthrough occurred in the system. At that point, copper, iron, lead and zinc were still completely removed, leaving no traces of the metals in the treated effluent. Copper uptake amounted to 16.11 mg g^{-1}, which is 2.23 times higher than the value obtained in a single ion laboratory column study. Arsenic and iron uptake amounted to 12.68 and 5.46 mg g^{-1}, respectively. The study showed that the two-stage treatment configuration is an ideal system for the simultaneous removal of copper and arsenic from low concentration GME, in addition to other heavy metals present at low concentrations.

Keywords: heavy metal; sorption; pilot system; gold mining effluent; coconut shell; iron-oxide-coated sand

8.1   Introduction

Heavy metals are found in wastewater discharged from industries such as mining, metal processing, electroplating, textile, tannery and petroleum refinery (Vilar et al., 2008; Izquierdo et al., 2010; Oguz and Ersoy, 2010; Kumar et al 2011; Martín-Lara et al., 2012). Heavy metals such as As, Cu, Pd, Zn, Fe, Cr, Ni and Hg are toxic (Kaczala et al., 2009; Liu et al., 2011; Saad et al., 2011) and therefore the discharge of untreated metal-laden effluent poses serious environmental challenges (Mondal, 2009; Benavente et al., 2011). Sorption using natural and agricultural materials has been proposed as a cost effective and environmentally friendly alternative to chemical treatment (Mohan and Chander, 2006; Sousa et al., 2009; Acheampong et al., 2010; 2011a; Fu and Wang, 2011).

Although there are many industrial processes using sorbents such as activated carbon, the development of biosorption processes is still mainly at the stage of laboratory studies in spite of enormous progress made over the last decade (Park et al., 2010). Most studies on continuous biosorption systems have focused mainly on the optimisation of operating conditions and breakthrough curves using synthetic metal solutions (Naja and Volesky, 2008; Muhamad et al., 2010; Sousa et al., 2010; Salamatinia et al., 2010; Kleinübing et al., 2011; Singh et al., 2011; Kleinübing et al., 2012; Ramírez Carmona et al., 2012; Izquierdo et al., 2012). Many of these studies have limited industrial application because industrial effluents are more complex, containing several metal ions and other contaminants (Singh et al., 2012). Singh et al. (2012) indicated that treatment of industrial effluent in continuous flow through systems allows the use of sorbents for multiple sorption and desorption applications. In this way, process sustainability is achieved through a substantial reduction in the fresh sorbent requirement, operational cost and solid waste materials needing disposal or containment (Park et al., 2010; Kumar et al., 2011). Little effort has been made on heavy metal removal from industrial effluents using agricultural and natural materials in continuous systems. For this reason, more research is needed at the pilot scale using real wastewater to demonstrate the applicability of a biosorption process to treat this type of wastewaters at the industrial scale.

The present work is a continuation of our previous studies, which showed the ability of coconut shell to sorb Cu(II) under batch (Acheampong et al., 2011a) and continuous flow through (Acheampong et al., 2013a) conditions, and its reusability during multiple sorption and desorption applications (Acheampong et al., 2013b). The SEM-EDX characterisation of coconut shell before and after sorption showed irregular surface texture of the biosorbent and the presence of K^+ and Mg^{2+} that were replaced by copper during sorption, while the FTIR analysis suggested the involvement of amino, amide, carboxylic, hydroxyl and carbonyl groups in metal binding (Acheampong et al., 2011a). Kinetic studies indicated that intraparticle diffusion controls the rate of sorption in the coconut shell biosorption of Cu(II); however film diffusion cannot be neglected, especially at the initial stage of sorption (Acheampong et al., 2011b).

In chapter 7, the biosorption of Cu(II) onto coconut shell was done at the laboratory scale with synthetic wastewater. This study evaluates the sorption efficiency of coconut shell (CS) and iron oxide coated sand (IOCS) for, respectively, copper (Cu) and arsenic (As) in gold mining effluent in a two-stage pilot fixed bed column continuous flow system operated from December 2011 to June 2012 with real GME (Temperature range: 25.7-27.6 ^0C, pH range: 7.6-8.1) from the AngloGold Ashanti mine (Obuasi, Ghana). Though the focus of this work was on the removal of Cu and As, the removal of other metals present in the gold mining effluent were considered as well.

8.2   Materials and methods

8.2.1   Sorbent preparation
The coconut shell (CS) used in this study was obtained from Kumasi (Ghana) and its Cu(II) biosorption properties were described in detail by Acheampong et al. (2011a). The material was washed with distilled water and dried at 105 ^0C for 24 h. The coconut shells were ground using a Peppink hammer mill. The biosorbent was then sieved to obtain the 0.5-1.4 mm fraction used throughout this study. The iron oxide-coated sand (IOCS) was obtained from the water treatment plant at Zwolle Engelse Werk (The Netherlands) as a spent product of the sand filtration process. The IOCS was used as received; with a particle size range of 1.0-3.0 mm. Coating analysis performed on the IOCS shows the presence of Mn (19.4 mg g^{-1}), Mg (0.81 mg g^{-1}), Ca (3.33 mg g^{-1}), Fe (151.85 mg g^{-1}) and Si (2.43 mg g^{-1}), among other constituents. In handling the IOCS before and during the experiments, we were careful not to destroy the iron oxide coatings on the sand particles. The physical characteristics of both sorbents used are presented in Table 1. SEM-EDX, FTIR and point of zero charge (PZC) analysis of the coconut shell prior to and after Cu(II) sorption were presented in Acheampong et al. (2011a).

8.2.2 Gold mining effluent

The GME (Table 8-1) was taken from the tailings pond at a location described in Acheampong et al. (2012c). The effluent was pumped into a 2500 L PVC storage tank (Polytank, GH-Pt2500) located at the tailings pond site. The tank was first cleaned thoroughly using a detergent, 10 % HNO_3, triple rinsed with distilled water, and finally triple rinsed with the GME. The effluent was kept in the storage tank over night to allow solid particles to settle at the bottom and the effluent to equilibrate. The supernatant was discharged by gravity into two 100 L plastic tanks (720011 Getrankesfass 100 L) and transported to the laboratory, located 25 km from the tailings pond, where the pilot plant was installed. The pH and the temperature of the effluent were checked prior to being fed to the sorption columns, but no pH adjustment was made. Table 8-1 shows the characteristics of the GME used for the experimental runs. Experimental runs I, II and III were conducted consecutively with GME collected on different dates from the same location but with different characteristics: the GME temperature, pH and metals concentrations were different for each run (Table 8-1). All GME samples were analysed three times and the values reported as the mean (± standard deviation).

Table 8-1: Characteristics of the gold mining effluent (untreated)

Parameters	Run I	Run II	Run III	Ghana EPA Standard
Temperature (^{0}C)	26.1 (±0.00)	25.7 (±0.02)	27.6 (±0.01)	40.0
pH	7.6 (±0.02)	7.7 (±0.01)	8.1 (±0.01)	6.0-9.0
Copper (mg L^{-1})	5.67 (±0.03)	6.21 (±0.01)	7.98 (±0.01)	1.00
Arsenic (mg L^{-1})	8.81 (±0.01)	8.72 (±0.02)	12.11 (±0.03)	0.20
Lead (mg L^{-1})	0.03 (±0.00)	0.05 (±0.00)	0.07 (±0.01)	0.10
Zinc (mg L^{-1})	0.01 (±0.00)	0.02 (±0.00)	0.01 (±0.00)	2.00
Iron (mg L^{-1})	1.92 (±0.01)	0.78 (±0.01)	2.16 (±0.02)	2.00
Cyanide (mg L^{-1})	1.56 (±0.03)	3.64 (±0.04)	2.58 (±0.01)	0.20

8.2.3 Pilot plant set-up

The experimental set-up (Fig. 8-1) was constructed out of two identical Perspex columns (each of 10 cm inner diameter and 200 cm long). Column I was packed with CS (particle size 0.5-1.4 mm) to give a total effective bed depth of 150 cm (5.17 kg of sorbent), while column II was packed with IOCS (particle size 1.0-3.0 mm) to give a total effective bed depth of 150 cm (10.06 kg sorbent). Each column used in this study had a packing volume of 11.8 L. The sorbent in each of the columns was placed in between a layer of glass beads at the top and a thin layer of sieve at the bottom. The sieve was to prevent sorbent particles from falling through the effluent outlet. Packing was done under wet conditions to prevent trapping of air within the bed. Table 8-3 shows the pilot plant operational parameters. Prior to the start of the experimental run, distilled water was passed through each of the columns until a clear liquid was obtained at the outlet. This was done to remove powdered sorbent particles from the columns.

The heavy metal removal efficiency was studied by pumping the GME from a 200 L plastic tank (L4506 Getrankesfass 200 L) into the two columns in series using a pump (Masterflex model 77521-57, Range: 1-100 rpm) equipped with two pump heads (Masterflex Easy-Load II, model 77201-60). Piping was done with PVC rubber tubing (Masterflex L/S 17 TYGON Tubing, Inner diameter = 6.4 mm, Outer diameter = 9.6 mm) with a flow range of 5-250 mL min^{-1}. The flow-rates were measured with a flow meter (OMEGA Direct Read Rotameter, model FLDW3210G, Maximum Flow Rate 250 mL min^{-1}) placed between the pump and each column (Figure 8-1). For regulating the liquid flow from the storage tanks, control valves were installed at the outlet of each storage tank as shown in Figure 8-1. Metal ion concentrations at the inlet of column I were measured prior to each run. Samples were collected from the outlet of each column twice daily for the first 5 days and then once a day for the remaining of the run time for metal ion analysis with AAS. Metal concentrations in the feed were checked regularly for fluctuations during the experiments; no significant variations were observed.

8.2.4 Experimental design

The pilot plant columns were operated at constant flow rate (150 mL min^{-1}) and fixed bed height (150 cm) during each run. Three experimental runs (I, II and III) were conducted with GME having different characteristics (Table 8-1). All the GME were collected from the same intake point but at different times. The reasons for running three different experiments with effluent having varied characteristics is to study the treatment plant response to fluctuations in metal concentrations as well as to verify the reproducibility of the results. All continuous operations with the pilot columns were performed under down-flow mode. The plant was operated until the dynamic transfer zone reached the end of the column (Figure 8-2). Until that time the effluent leaving the column had no trace of the sorbate in it. When the transfer zone reaches the column end, the sorbate concentration in the effluent starts to gradually increase and, for practical purposes, the working life of the column is over: the "breakthrough point" is reached, marking the usable column "service time". The operation of the plant was stopped when the outlet copper (Cu) and arsenic (As) concentrations exceeded the breakthrough concentration, defined as the regulatory limit (Table 8-1) of the Ghana EPA for industrial effluent discharge (Ghana EPA, 2010). Each sample analysis was carried out in triplicate and results reported are average with ± 3% standard deviation error. The pilot plant was operated December 2011 to June 2012.

Table 8-2: Pilot plant operational parameters

Parameters	Column I (CS)	Column II (IOCS)
Column height (cm)	200	200
Bed height (cm)	150	150
Mass of sorbent used (kg)	5.17	10.06
Column diameter (cm)	10	10
Particle diameter for packing (mm)	0.5-1.4	1.0-3.0
Flow rate (mL min^{-1})	150	150

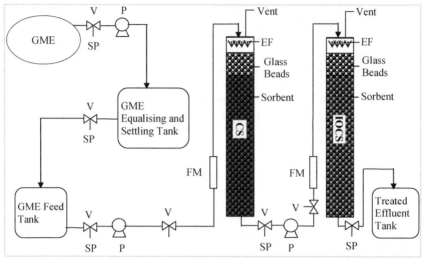

Figure 8-1: Schematics diagram of the two-stage pilot plant set-up. CS: Coconut shell; IOCS: Iron Oxide Coated Sand; EF: Effluent Distributor; P: Pump; SP: Sampling Point; V: Valve; GME: Gold Mining Effluent; FM: Flow Meter.

8.2.5   Analytical techniques

The concentrations of copper, lead, iron and zinc in the samples were measured using an atomic absorption spectrometer (AAS-Air Acetylene Flame, Perkin Elmer, model AAnalist200) at wavelengths of 324.80, 283.31248.33 and 213.86, respectively. The arsenic concentration in the samples was determined by an atomic absorption spectrometer (AAS-Graphite Furnace, Solar MQZe GF95). The detection limit of the AAS for copper, lead, iron, zinc and arsenic are 0.01, 0.1, 0.05, 0.02 and 0.001 mg L^{-1}, respectively. The quality assurance procedure followed for the metal analysis was described in detail by Acheampong et al. (2011a) (Chapter 4).

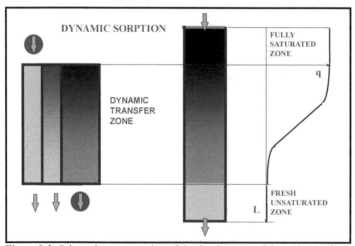

Figure 8-2: Schematic representation of the development of dynamic transfer zone in the continuous-flow fixed bed sorption column

Temperature and pH were determined by a multi-parameter ion specific meter (Hanna instrument, combo). The pH meter was calibrated using buffer solutions with pH values of 4, 7 and 10.

8.2.6   Calculations

The sorption capacity was calculated from Eq. 1 (Kumar et al., 2011):

$$q = \frac{t \times Q \times C_{in}}{M_s} \tag{8-1}$$

where t is the treatment time (min), Q is the flow rate (mL min^{-1}), C$_i$ is the initial concentration (mg L^{-1}) and M$_s$ is the dry mass of sorbent (g).

The total effluent volume treated, V (mL), was calculated from Eq. 3 (Futalan et al., 2011):

$$V = Qt \tag{8-2}$$

The metal removal efficiency of the treatment system at any time during operation was calculated based on the inlet and the outlet effluent concentrations as follows:

$$removal\ efficiency\ (\%) = \left[\frac{\left(C_{in} - C_{out} \right)}{C_{in}} \right] \times 100 \tag{8-3}$$

where, C$_{in}$ = Influent concentration (mg L^{-1}) and C$_{out}$ = Outlet effluent concentration (mg L^{-1})

8.3   Results

8.3.1   Metal removal in CS and IOCS columns

The metal removal efficiency of the two-stage pilot plant is given in Figures 8-3 and 8-4. In these figures, the outlet As and Fe effluent concentration is plotted against the treatment time. Copper and iron removal in column I (packed with CS) is shown in Figure 8-3. Since these were completely removed in the columns, no Cu and Fe effluent values were reported for column II. Similarly, as no arsenic removal occurred in column I, Figure 8-4 shows arsenic removal in column II (IOCS) only. The maximum Cu and Fe concentrations in the treated effluent (Figure 8-3) were recorded at As breakthrough, when the As concentration was 0.2 mg L^{-1}, which is the permissible limit for As for the discharge of industrial effluent (Ghana EPA, 2010).

Figure 8-4 shows that at the end of the treatment process (at As breakthrough with an As concentration of 0.2 mg L^{-1}), no traces of other heavy metals investigated (Cu, Fe. Pb and Zn) were found in the treated effluent from column II (data not shown). The removal efficiency and uptake for the individual columns are presented in Table 8-3. The time for performing the experiments till breakthrough in runs I, II, and III was 1608, 1464 and 1296 h, respectively (Table 8-3). At the end of each of the treatment periods (runs I, II and III), only As was detected in the final effluent.

8.3.2   Kinetics of metal uptake and removal efficiency in the treatment plant

The overall metal uptake and removal efficiency for runs I, II and III (for the entire process) is shown in Table 8-4. The overall metal uptake as a function of treatment time for runs I, II and III is presented in Figure 8-5.

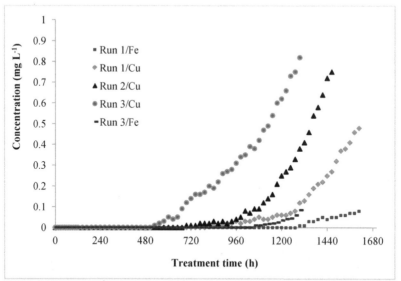

Figure 8-3: Amount of copper (Cu) and iron (Fe) remaining in the treated effluent from column I in runs I, and 2. Bed height = 150 cm, flow rate = 150 mL min^{-1}, inlet concentrations: run I (Cu = 5.65 mg L^{-1}, Fe = 1.92 mg L^{-1}), run 2 (Cu = 6.21 mg L^{-1}, Fe = 0.78 mg L^{-1}) and run 3 (Cu = 7.98 mg L^{-1}, Fe = 2.11 mg L^{-1})

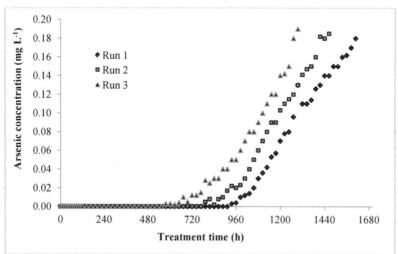

Figure 8-4: Amount of arsenic (As) remaining in the treated effluent from column II. Bed height = 150 cm, flow rate = 150 mL min^{-1}, inlet concentrations: run I = 8.81.65 mg L^{-1}, run II = 8.72 mg L^{-1}, and run III = 12.11 mg L^{-1}

A steady increase in Cu/Fe uptake by CS and As uptake by IOC during the treatment period was observed in, respectively, the CS and IOCS columns (Figure 8-5). The results showed that upon As breakthrough in column II, the two-stage pilot treatment plant still removed 100% of the Cu and Fe entering the system (Table 8-4).

Table 8-3: Metal uptake and removal efficiency for the difference runs (bed height = 150 cm, flow rate = 150 mL min^{-1})

Parameters	Metal	Column I			Column II		
		Run I	Run II	Run III	Run I	Run II	Run III
Treatment time, t (h)	Cu	1,608	1,464	1,296			
	Fe	1,608	1,464	1,296			
	As				1,608	1,464	1,296
Volume treated, V (m^3)	Cu	14.5	13.2	11.7			
	Fe	14.4	13.2	11.7			
	As				14.5	13.2	11.7
Uptake, q (mg g^{-1})	Cu	15.88 (±0.02)	15.83 (±0.01)	18.01 (±0.03)			
	Fe	5.38 (±0.01)	1.99 (±0.001)	4.87 (±0.02)			
	As				12.68 (±0.02)	11.43 (±0.03)	14.05 (±0.03)
Removal, R (%)	Cu	91.5 (±0.02)	87.9 (±0.03)	89.7 (±0.02)			
	Fe	95.3 (±0.03)	100 (±0.01)	93.8 (±0.01)			
	As				97.7 (±0.03)	97.7 (±0.003)	98.3 (±0.02)

Table 8-4: Overall plant performance (columns I and II) for copper, iron and arsenic removal

Parameter	Metal	Overall plant performance		
		Run I	Run II	Run III
Uptake, q (mg g^{-1})	Cu	16.11 (±0.01)	16.09 (±0.02)	18.34 (±0.01)
	Fe	5.46 (±0.01)	2.02 (±0.00)	4.96 (±0.01)
	As	12.68 (±0.02)	11.43 (±0.03)	14.05 (±0.03)
Removal, R (%)	Cu	100.0 (±0.01)	100.0 (±0.01)	100.0 (±0.02)
	Fe	100.0 (±0.00)	100.0 (±0.01)	100.0 (±0.01)
	As	97.7 (±0.03)	97.7 (±0.03)	98.3 (±0.02)

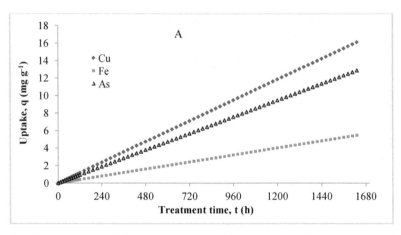

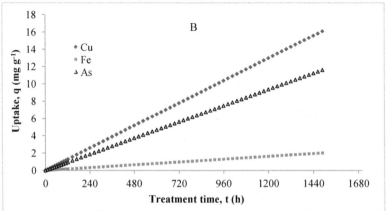

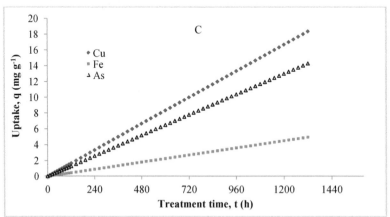

Figure 8-5: Kinetics of copper, iron and arsenic uptake in the treatment plant (columns I + II) for run I (A), run II (B), run III (C)

8.4   Discussion

8.4.1   Performance of the two-stage pilot plant treating the gold mining effluent

8.4.1.1   Arsenic removal

This study shows that a two-stage pilot plant consisting of a coconut shell (CS) and an iron oxide-coated sand (IOCS) packed column was capable of removing heavy metals from gold mining effluent (GME). The combination of a CS and IOCS columns in a two-stage treatment plant thus offers the possibility of removing several metals at the same time, in a cost effective manner. The operational run time was determined by As breakthrough, as the later was accumulating in the IOCS effluent when Cu, Fe, Pb and Zn were still 100% removed.

Arsenic adsorption is influenced by the solution pH. $H_2AsO_4^-$ and $HAsO_4^{2-}$ are the predominant As(V) species at pH below 10 (Zhang et al., 2011). Many studies suggest that increasing the pH decreases As(V) adsorption on iron and iron-containing adsorbents (Jang et al., 2006; Zhang et al., 2007), which is typical for anionic adsorption. These studies also indicate that the effect of pH on As(III) adsorption on iron-containing adsorbents is different from that of As(V) since H_3AsO_3 is the predominant dissolved As(III) species at pH below 9.2. As(III) exists in non-ionic (H_3AsO_3) and anionic ($H_2AsO_3^-$ and $HAsO_3^{2-}$) form in the pH range of 2-7 and 7.5-9, respectively (Dhoble et al., 2011). As(III) removal was optimal at pH 7.7 using Fe-Mn adsorbent (Benavente et al., 2011). They indicated that the As(III) was first oxidised to As(V) and then removed by the iron-based adsorbent. Iron-based adsorbents are effective in removing arsenic from aqueous medium in the pH range of 6.5-8.5 (Gu et al., 2005). Similarly, the iron-oxide-coated sand used in this study was able to remove arsenic from the gold mining effluent studied to meet the Ghana EPA standard value (0.2 mg L^{-1}).

Due to its high arsenic uptake and low cost, IOCS represents an attractive adsorbent for arsenic removal from wastewater. In the pH range of the GME studied (Table 8-1), As exists mainly as As(III). According to the literature, As(III) is more toxic and mobile than As(V) (Thirunavukkarasu et al., 2003; Gupta et al., 2005; Ngantcha et al., 2011) and more difficult to remove by many treatment technologies (Zhang et al., 2007). The effectiveness of adsorbents such as IOCS and manganese oxide-coated alumina (MOCA) in removing As from solutions was explained by the removal mechanism, which involves the oxidation of As(III) to As(V) and adsorption of the As(V) onto the sorbent surface (Thirunavukkarasu et al., 2003; Maliyekkal et al., 2009). Zhang et al. (2003; 2007) further suggested that in addition to the oxidation of As(III) to As(V) by the oxide coated adsorbent, fresh adsorption sites for As adsorption were created at the solid surface during As(III) oxidation, resulting in an increased As removal. This argument was supported by Deschamps et al. (2003; 2005), who indicated that oxide-coated minerals are effective oxidants for the transformation of As(III) to As(V), allowing the uptake of large quantities of arsenic. The long bed height (150 m) used in this study ensured a longer contact time and availability of more sorption sites for arsenic uptake compared to smaller laboratory based columns (Zhang et al., 2007).

8.4.1.2   Copper removal

The removal of Cu and Fe in column I, packed with CS (Figure 8-1), shows the potential of CS to sorb divalent heavy metals from aqueous solutions. Although Pb and Zn concentrations in the initial GME (Table 8-1) were very low and therefore of no threat to the environment, their removal in column I, as evident by their absence from the treated effluent (data not shown), is significant because it suggests that CS columns are capable of removing elevated concentration of divalent heavy metals. Benavent et al. (2011) reported the treatment of GME using chitosan and found that the removal of Cu, Pb and Zn was not affected by the presence of other metals because they were present at low concentrations.

In treating electroplating wastewater using sugar cane bagasse, Sousa et al. (2009) reported a considerable decrease in adsorption of zinc and nickel, probably due to electrostatic repulsive forces between the cationic sugar cane bagasse surface and the metal ions. In contrast, the CS used in this study was negatively charged at the GME pH due to the lower point of zero charge pH of 6.5 (Acheampong et al., 2011a), suggesting that the removal of the positively charged heavy metals from the GME could be enhanced by electrostatic attraction forces between the sorbent and the sorbate.

The tailings dam wastewater used in this study also contained 2.58 mg L^{-1} cyanide (Table 8-1). In the presence of cyanide, metals predominantly exist in uncomplexed form (Sousa et al., 2009). Benavete et al. (2011) indicated that the presence of cyanide had a negative impact on the sorption process, but not to the extent of affecting the removal of copper significantly, as evident from the low copper concentration (< 0.01 mg L^{-1}) in their treated effluent. Similar to the findings of Benavete et al. (2011), we observed that neither the presence of cyanide nor other heavy metals inhibited the removal of copper from the GME (Table 8-4).

The simultaneous removal of Cu, Fe, Pb and Zn in the CS packed column was expected to reduce the Cu uptake of 7.25 mg g^{-1} obtained in a single ion laboratory column study (Acheampong et al., 2013a) (Chapter 7). On the contrary, a much higher uptake of Cu was recorded in the three runs (Table 8-4), indicating a positive impact of an increased bed height on the metal uptake or co-sorption of Cu with the other metals. This may be attributed to the increase in the CS mass and the corresponding increase in the available sorption sites for metal uptake (Kumar et al., 2011). A lower Cu concentration, a lower flow rate and higher bed height is needed for an optimum performance of the biosorption treatment system (Acheampong et al., 2013a). In this study, the Cu concentration in the GME was lower than the 10 mg L^{-1} used in the laboratory column studies (Acheampong et al., 2013a) (Chapter 7). Therefore, a substantial increase in the CS bed height produced a significant increase in Cu uptake, even in the presence of other metals.

8.4.2   Implication for industrial application

The successful application of the treatment system in removing heavy metals from the GME shown in this study is an indication of the suitability of the system for treating large volumes of GME having low metal concentrations. The pilot study thus has a positive implication for industrial applications. According to Volesky (2003), virtually unlimited scale-up of the biosorption process is accomplished by using batteries of multiple columns that work in parallel and/ or series to optimise the performance of the process. Thus, increasing GME volumes can be treated by using larger and bigger diameter columns in the two-stage treatment plant (Figure 8-1). For an effective design, cylindrical sorption columns (used in this study) must exceed 1.5 m in diameter and 5 m in height (Volesky, 2003).

The long bed height used in this study increased the time required to attain the breakthrough point, leading to an increased GME volume treated. In contrast, Singh et al. (2012) found that the Cu(II) and Pb(II) sorption capacity of *Spirogyra* biomass did not increase significantly when the bed height was increased. Although they did not assign any specific reason for this, they indicated that the process could ensure the treatment of large effluent volumes. Since there was a significant increase in uptake as a consequence of the increased CS bed height used in this study (Table 8-4), an increased treatment time and volume of GME treated was apparent. This is important because the columns can be operated for a much longer time before breakthrough is reached, reducing the need for a frequent sorbent replacement or regeneration.

When the system attains breakthrough, the column is shut down and the flow is diverted into a second stand-by column packed with fresh sorbent. The saturated column is then appropriately processed (regenerated, washed) to prepare it for another run. Based on the performance data from this study, 1000 kg of CS and 1950 kg of IOCS can treat approximately 2800 m^3 of GME. The gold processing plant discharges an average of 450 m^3 h^{-1} of wastewater into the tailings dam, corresponding to an average of 12000 m^3 discharged into the dam in a 24 h working day. Therefore, approximately 4000 kg of CS and 7800 kg of IOCS are needed per day to treat the effluent produced. The number of columns needed per day would depend on the preferred configuration, column dimension and space availability.

The CS can be used in a cyclic sorption and desorption application without loss of uptake capacity (Acheampong et al., 2013b) (Chapter 5). According to Acheampong et al. (2013b), the metal loaded-CS can be regenerated with 0.05M HCl over eight sorption/desorption cycles without loss of capacity and structural damage to the CS. Therefore, the requirement for fresh sorbent is reduced, making the process more cost effective and sustainable. To ensure uninterrupted operation at industrial level, integration of sorbent production and processing into the treatment system is vital. Regeneration and recovery of Cu and other metals from the CS, and the disposal of the spent sorbent are important from environmental, economic and public health point of view. For the present system, incineration of the spent CS after a Cu recovery step will lead to the production of energy as resource. The ash produced from the incineration together with the spent IOCS will still need to be disposed off at an engineered landfill site, eventually after Cu bioleaching and recovery step.

8.5   Conclusions

This study showed that a two-stage pilot system of columns packed with CS and IOCS allows the simultaneous removal of heavy metals (Cu, Pb, Fe, Zn and As) from GME to meet the regulatory discharge standards for each metal present. At constant bed height and effluent flow rate, GME containing lower metal concentrations (run I) allowed for a longer time (1,608 h) to achieve breakthrough, leading to the treatment of large volumes of effluent. The ability of the pilot-scale two-stage sorption plant to efficiently remove heavy metals from the GME makes it suitable for large-scale applications. The capacity of the treatment plant can be increased by using batteries of multiple columns configured in parallel and/or series.

8.6   Acknowledgements

The authors acknowledge funding from the Netherlands Government under the Netherlands Fellowship Programme (NFP), NUFFIC award (2009-2013), Project Number: 32022513. We also acknowledge funding from the Staff Development and Postgraduate Scholarship Scheme (Kumasi Polytechnic, Ghana) and the UNESCO-IHE Partner Research Fund, UPaRF III research project PRBRAMD (No. 101014). We are grateful to AngloGold Ashanti (Obuasi mine, Ghana) for providing facilities.

8.7   References

Acheampong, M.A., Pereira, J.P.C., Meulepas, R.J.W. and Lens, P.N.L., 2011a. Biosorption of Cu(II) onto agricultural materials from tropical regions. Journal of Chemical Technology & Biotechnology 86, 1184-1194.

Acheampong, M.A., Pereira, J.P.C., Meulepas, R.J.W. and Lens, P.N.L., 2011b. Kinetics modelling of Cu(II) biosorption on to coconut shell and *Moringa oleifera* seeds from tropical regions. Environmental Technology 33, 409-417.

Acheampong, M.A., Pakshirajan, K., Annachhatre, A.P., Lens, P.N.L., 2013a. Removal of Cu(II) by biosorption onto coconut shell in fixed bed column systems, Journal of Industrial and Engineering Chemistry 19, 841-848.

Acheampong, M.A., Dapcic, A.D., Yeh, D., Lens, P.N.L., 2013b. Cyclic sorption and desorption of Cu(II) onto coconut shell and iron oxide coated sand, Separation Science and Technology. (DOI:10.1080/01496395.2013.809362) (In press).

Acheampong, M.A., Pakshirajan, Lens, P.N.L., 2013c. Assessment of the effluent quality from a gold mining industry in Ghana, Environmental Science and pollution research 20, 3799-3811.

Benavente, M., Moreno, L. and Martinez, J., 2011. Sorption of heavy metals from gold mining wastewater using chitosan. Journal of the Taiwan Institute of Chemical Engineers 42, 976-988.

Deschamps, E., Ciminelli, V.S.T. and Höll, W.H., 2005. Removal of As(III) and As(V) from water using a natural Fe and Mn enriched sample. Water Research 39, 5212-5220.

Deschamps, E., Ciminelli, V.S.T., Weidler, P.G. and Ramos, A.Y., 2003. Arsenic sorption onto soils enriched in Mn and Fe minerals. Clays and Clay Minerals 51, 197-204.

Fourest, E. and Roux, J.-C., 1992. Heavy metal biosorption by fungal mycelial by-products: mechanisms and influence of pH. Applied Microbiology and Biotechnology 37, 399-403.

Ghana EPA, 2010. Environmental performance rating and disclosure: report on the performance of mining and manufacturing companies. Environmental Protection Agency, Accra, Ghana.

Izquierdo, M., Gabaldón, C., Marzal, P. and Álvarez-Hornos, F.J., 2010. Modeling of copper fixed-bed biosorption from wastewater by Posidonia oceanica. Bioresource Technology 101, 510-517.

Izquierdo, M., Marzal, P., Gabaldón, C., Silvetti, M. and Castaldi, P., 2012. Study of the Interaction Mechanism in the Biosorption of Copper(II) Ions onto *Posidonia oceanica* and Peat. CLEAN – Soil, Air, Water 40, 428-437.

Kaczala, F., Marques, M. and Hogland, W., 2009. Lead and vanadium removal from a real industrial wastewater by gravitational settling/sedimentation and sorption onto Pinus sylvestris sawdust. Bioresource Technology 100, 235-243.

Kleinübing, S.J., da Silva, E.A., da Silva, M.G.C. and Guibal, E., 2011. Equilibrium of Cu(II) and Ni(II) biosorption by marine alga *Sargassum filipendula* in a dynamic system: Competitiveness and selectivity. Bioresource Technology 102, 4610-4617.

Kleinübing, S.J., Guibal, E., da Silva, E.A. and da Silva, M.G.C., 2012. Copper and nickel competitive biosorption simulation from single and binary systems by *Sargassum filipendula*. Chemical Engineering Journal 184, 16-22.

Kumar, R., Bhatia, D., Singh, R., Rani, S. and Bishnoi, N.R., 2011. Sorption of heavy metals from electroplating effluent using immobilized biomass *Trichoderma viride* in a continuous packed-bed column. International Biodeterioration & Biodegradation 65, 1133-1139.

Liu, C.C. et al., 2011. Biosorption of chromium, copper and zinc on rice wine processing waste sludge in fixed bed. Desalination 267, 20-24.

Maliyekkal, S.M., Philip, L. and Pradeep, T., 2009. As(III) removal from drinking water using manganese oxide-coated-alumina: Performance evaluation and mechanistic details of surface binding. Chemical Engineering Journal 153, 101-107.

Martín-Lara, M.A., Blázquez, G., Ronda, A., Rodríguez, I.L. Calero, M., 2012. Multiple biosorption–desorption cycles in a fixed-bed column for Pb(II) removal by acid-treated olive stone, Journal of Industrial and Engineering Chemistry 18, 1006-1012.

Mohan, D. and Chander, S., 2006. Removal and recovery of metal ions from acid mine drainage using lignite—A low cost sorbent. Journal of Hazardous Materials 137, 1545-1553.

Muhamad, H., Doan, H. and Lohi, A., 2010. Batch and continuous fixed-bed column biosorption of Cd^{2+} and Cu^{2+}. Chemical Engineering Journal 158, 369-377.

Naja, G. and Volesky, B., 2008. Optimization of a Biosorption Column Performance. Environmental Science & Technology 42, 5622-5629.

Oguz, E. and Ersoy, M., 2010. Removal of Cu^{2+} from aqueous solution by adsorption in a fixed bed column and Neural Network Modelling. Chemical Engineering Journal 164, 56-62.

Park, D., Yun, Y.-S. and Park, J., 2010. The past, present, and future trends of biosorption. Biotechnology and Bioprocess Engineering 15, 86-102.

Ramírez Carmona, M.E., Pereira da Silva, M.A., Ferreira Leite, S.G., Vasco Echeverri, O.H. and Ocampo-López, C., 2012. Packed bed redistribution system for Cr(III) and Cr(VI) biosorption by Saccharomyces cerevisiae. Journal of the Taiwan Institute of Chemical Engineers 43, 428-432.

Saad, D.M., Cukrowska, E.M. and Tutu, H., 2011. Development and application of cross-linked polyethylenimine for trace metal and metalloid removal from mining and industrial wastewaters. Toxicological & Environmental Chemistry 93, 914-924.

Salamatinia, B., Kamaruddin, A.H. and Abdullah, A.Z., 2010. Regeneration and reuse of spent NaOH-treated oil palm frond for copper and zinc removal from wastewater. Chemical Engineering Journal 156, 141-145.

Singh, P., Bajpai, J., Bajpai, A.K. and Shrivastava, R.B., 2011. Fixed-Bed Studies on Removal of Arsenic from Simulated Aqueous Solutions Using Chitosan Nanoparticles. Bioremediation Journal 15, 148-156.

Sousa, F.W. et al., 2009. Evaluation of a low-cost adsorbent for removal of toxic metal ions from wastewater of an electroplating factory. Journal of Environmental Management 90, 3340-3344.

Sousa, F.W. et al., 2010. Green coconut shells applied as adsorbent for removal of toxic metal ions using fixed-bed column technology. Journal of Environmental Management 91, 1634-1640.

Thirunavukkarasu, O.S., Viraraghavan, T. and Subramanian, K.S., 2003. Arsenic Removal from Drinking Water using Iron Oxide-Coated Sand. Water, Air, and Soil Pollution 142, 95-111.

Vilar, V.J.P., Loureiro, J.M., Botelho, C.M.S. and Boaventura, R.A.R., 2008. Continuous biosorption of Pb/Cu and Pb/Cd in fixed-bed column using algae Gelidium and granulated agar extraction algal waste. Journal of Hazardous Materials 154, 1173-1182.

Volesky, B., 2003. Sorption and Biosorption. BV Sorbex, Inc.,Montreal, Canada, pp. 179-190.

Zhang, G., Qu, J., Liu, H., Liu, R. and Wu, R., 2007. Preparation and evaluation of a novel Fe–Mn binary oxide adsorbent for effective arsenite removal. Water Research 41, 1921-1928.

Chapter 9

9 General Discussion

The modified version of this chapter will be published as:

Acheampong, M.A., Lens, P.N.L., 2013. Biosorption applications for environmental technological processes using agricultural biomaterials. Trends in Environmental Biotechnology (In preparation).

9.1   Introduction

Sorption techniques for heavy metal removal have been studied extensively during the last few decades. However, the application of the technology at the pilot level using the real wastewater has not been given the much needed attention. Industrial application of sorption may not be possible if efforts are not directed toward field testing at the pilot scale. The application aspect is what makes the research and development work in the relatively new area exciting. Therefore, the research on sorption techniques for the sustainable treatment of gold mining wastewater described in this thesis focused on the process development that ultimately results in the industrial scale use of the technology. This chapter discusses the progress made within this research leading to the successful treatment of real gold mining effluent in pilot fixed-bed columns. In addition, recommendations for further research are given.

9.2   Gold mining effluent characterisation

The study of a wastewater stream's characteristics helps to identify contaminants present in the wastewater and provides information on the treatment required to meet the regulatory limits for discharge. The results of the research conducted with the process effluent and the Sansu tailings dam wastewater quality of the AngloGold Ashanti mining company in Ghana revealed high levels of contamination (Chapter 3). Almost all the parameters measured were found to exceed the permissible limit set by the Ghana EPA and the WHO guideline values (Chapter 3). Though the characterisation results showed the presence of heavy metals and other physicochemical pollutants, copper, arsenic and cyanide were identified as the most toxic constituents. The wastewater from the Sansu tailings dam cannot be discharged into the environment without prior treatment. Any treatment technology for this effluent must focus on these contaminants. Sorption using agricultural and plant materials or natural sorbents could be a low cost option for treating this effluent.

9.3   Sorption techniques for sustainable gold mining wastewater treatment

Screening of the four locally obtained agricultural materials for their sorption capacity constituted an important step in the sorption process (Chapter 4). The performance of the coconut husk and sawdust was low, and neither of them was able to reach a metal uptake capacity exceeding 1 mg g^{-1}. The coconut shell and the *Moringa oleifera* seeds removed Cu(II) efficiently, accomplishing low equilibrium concentrations in solution. However, for industrial application, only coconut shell can be considered as an important low-cost sorbent for Cu(II) removal due to its high affinity for copper ions. The sorption of Cu(II) onto coconut shell was best described by the Freundlich isotherm model which takes surface roughness into account.

The adsorption capacity is strongly influenced by the surface structures of carbon-oxygen-hydrogen functional groups and surface behaviour of carbon (Rahman and Islam, 2009; Sarioglu et al., 2009). A characterisation study was performed on the coconut shell and *Moringa oleifera* seeds to establish the removal mechanism of copper. The FTIR analysis results (Chapter 4) indicated that hydroxyl and carboxylic functional groups were involved in Cu(II) removal by coconut shell, coconut husk and sawdust through the ion exchange mechanism and that mainly amino functional groups were involved in Cu(II) removal by *Moringa oleifera* seeds (Chapter 4). The EDX analysis results suggest that the release of magnesium and potassium ions which were initially fixed onto the coconut shell was followed simultaneously by sorption of Cu(II). Also magnesium and potassium present on the *Moringa oleifera* seeds were replaced after sorption of Cu(II). The exchange between Mg(II), K(I) and Cu(II) on *Moringa oleifera* seeds and between K(I) and Cu(II) on coconut shell indicated the involvement of ion exchange mechanisms in the sorption process for the removal of copper by both sorbents. Furthermore, the PZC results (Chapter 4) suggest that electrostatic forces played a role in the sorption of Cu(II) in this study.

To study the desorption ability of both CS and IOCS, five different desorption solutions were screened under batch conditions using the loaded sorbent (all at a 0.2 M concentration): hydrochloric acid (HCl), ethylenediaminetetraacetic acid (EDTA), sodium hydroxide (NaOH), sodium acetate ($C_2H_3NaO_2$) and calcium nitrate ($Ca(NO_3)_2$) (Chapter 5). Desorption of sorbed metals from the loaded sorbent enables reuse of the sorbent, and recovery and/or containment of the sorbed metals from a highly concentrated eluant stream (Gupta and Rastogi, 2008; Gadd, 2009). The study showed that CS can be used in sorption and desorption cycles to treat water containing dissolved Cu(II) ions up to 10 mg L^{-1} using HCl (Chapter 5). HCl, even after eight cycles, does not affect CS's ability to sorb Cu(II) ions nor does it deteriorate its particle structure. Therefore, the sorbent does not need to be replaced every single sorption step and the application of such a sorption/desorption cycle decreases process costs. In addition, rapid sorption and desorption times (60 minutes) add to its applicability to treat larger volumes of wastewater. Though IOCS exhibited a high capacity to sorb Cu(II) ions, it cannot be used for cyclical sorption/desorption applications due to the structural damage caused by the desorbents investigated (Chapter 5).

In order to gain a better understanding of the sorption process, various kinetic models were used to test the experimental data. The overall sorption process may be controlled by film diffusion, intraparticle diffusion or sorption on the surface (Febrianto et al., 2009). The results indicate that kinetic data were best described by the pseudo second-order model with correlation coefficient (R^2) of 0.9974 and 0.9958 for the coconut shell and *Moringa oleifera* seeds, respectively (Chapter 6). On the basis of the correlation coefficients, the sorption of Cu(II) by the two sorbents follows a second-order reaction pathway. The values of the mass transfer coefficients obtained for coconut shell ($\beta_l = 1.2106 \times 10^{-3}$ cm s^{-1}) and *Moringa oleifera* seeds ($\beta_l = 8.965 \times 10^{-4}$ cm s^{-1}) indicate that the transport of Cu(II) from the bulk liquid to the solid phase was quite fast for both materials investigated (Chapter 6). The results indicate that intraparticle diffusion controls the rate of sorption in this study; however film diffusion cannot be neglected, especially at the initial stage of sorption.

The continuous removal of Cu(II) by coconut shell in a fixed-bed column was done to determine the optimum operating conditions (Chapter 7). The evaluation of the sorption performance in a continuously operated column was necessary because the sorbent uptake capacity is more efficiently utilised than in a completely mixed system (Chandra-Sekhar et al., 2003; Vilar et al., 2008). The fixed-bed sorption system was found to perform better for Cu(II) uptake by coconut shell at a lower Cu(II) inlet concentration, lower feed flow rate and higher coconut shell bed-depth (Chapter 7).

Under these optimum conditions, the service time to breakthrough and Cu(II) concentration were 58 h and 0.8 mg L^{-1}, respectively, after which the Cu(II) concentration in the effluent exceeded the 1 mg L^{-1} discharge limit set by the Ghana Environmental Protection Agency (Ghana EPA, 2010).

According to Singh et al. (2009) and Han et al. (2006a), the increase in the number of binding sites resulting from the increase in the adsorption surface area accounted for the increased uptake of Cu(II). The column performed well at the lowest flow rate because at the lower flow rate, the residence time of the feed solution was increased, allowing Cu(II) ions to diffuse into the pores of the coconut shell by means of intra-particle diffusion (Luo et al., 2011a). Among the various breakthrough models tested in this study (Chapter 7), the Clark model could be considered a more refined model as it involves both the mass transfer and equilibrium adsorption in predicting the breakthrough phenomena. From the modelling results, the design of a continuous fixed-bed column treatment system for copper laden wastewater can be achieved using the BDST, Yoon-Nelson and Clark breakthrough models.

The results of the pilot study on the gold mining effluent (GME) of the AngloGold Ashanti mine (Obuasi, Ghana) shows that coconut shell (CS) and iron oxide-coated sand (IOCS) packed column systems were capable of removing heavy metals from gold mining effluent (Chapter 8). Singh et al. (2012) indicated that treatment of industrial effluent in continuous flow through systems allows the use of sorbents for multiple sorption and desorption applications. In this way, process sustainability is achieved through the substantial reduction in the fresh sorbent requirement, operational cost and solid waste materials needing disposal or containment (Park et al., 2010; Kumar et al., 2011). While both the CS and IOCS have affinity for Cu, only the IOCS was able to remove As from the effluent. The combination of CS and IOCS in a single treatment plant offers the possibility of removing several metals at the same time, and in a cost effective manner.

9.4   Implication for industrial applications

The pilot study has a positive implication for industrial applications. The successful application of the treatment system in removing heavy metals from the GME is an indication of the suitability of the system for treating large volumes of GME having low metal concentrations (Chapter 8). According to Volesky (2003), virtually unlimited scale-up of the process is accomplished by using batteries of multiple columns that work in parallel and/or series to optimise the performance of the process. However, cylindrical sorption columns (used in this study) do not typically exceed 1.5 m in diameter and 5 m in height (Volesky, 2003). When the system attains the breakthrough point, the column is shut down and the flow is diverted into a second stand-by column packed with fresh sorbent. The saturated column is then appropriately processed (regenerated, washed) to prepare it for another run. Based on the performance data from this study (Chapter 8), 1000 kg of CS and 1950 kg of IOCS can treat approximately 2800 m^3 of GME.

The CS can be used in a cyclic sorption and desorption application without loss of uptake capacity (Chapter 5). Therefore, the requirement for fresh sorbent is reduced, making the process cost effective and sustainable. To ensure uninterrupted operation at industrial level, integration of sorbent production and processing into the treatment system is vital. Regeneration and recovery of Cu and other metals from the CS, and the disposal of the spent sorbent are important from environmental, economic and public health point of view. For the present system, incineration of the spent CS will lead to the production of energy as resource. The ash produced from the incineration together with the spent IOCS can then be disposed off at an engineered landfill site.

9.5 Further research

9.5.1 Coconut shell (CS) processing and formulation

Biomass processing and formulation presents quite a different type of activity when compared to studying and exploiting its sorbent properties (Volesky, 2003). This PhD research has established the sorption behaviour of the native CS in fixed-bed column systems. However, there is the need to process the CS into ready to use sorbent granules. Therefore, further research is required to establish the effectiveness of CS granulation and reinforcement. In order to make CS granules, the CS needs to be immobilised, reinforced - if required, and formed into suitable solid particles (granulated).

Most of the technologies used for particle making (granulation) have been reasonably well developed for many different types of materials. However, the special nature and widely varied chemistry associated with different biomass types requires special efforts to obtain the desired granule properties. Granulation procedures could differ substantially depending whether dry or wet biomass is to be processed. Consequently, there is the need to develop and operate a CS processing pilot plant alongside the pilot plant sorption unit. In this way, the technology can easily be transferred into an industrial scale process.

9.5.2 Effect of chemical treatment on sorption performance of coconut shell (CS) for copper

When the (bio) sorbent undergoes chemical treatment, there are always chances that its performance may suffer (Volesky, 2003). Further research is required to investigate if modification of the CS material, either chemically or physically, can considerably improve the Cu(II) uptake capacity of the CS without compromising its regeneration and reuse potential.

9.5.3 Metal recovery

Metal "recovery" is a down-stream consideration added to the metal "removal" aspect which is driven predominantly by environmental "detoxification" considerations. The sorption and desorption processes offered the opportunity for an economical recovery of copper from the concentrated solution. The feasibility of the recovery operation depends, to a large degree, on the type of recovery procedures used. The optimisation of the cost effective recovery of copper from the concentrated solution obtained during the regeneration of the loaded sorbent, e.g. through electrochemical techniques such as electro-winning, needs to be investigated.

9.5.4   Disposal of the spent sorbents

The spent sorbents generated at the end of the process must be properly and safely disposed. Various disposal options that are in conformity with local solid waste management practice needs to be investigated and the appropriate option selected.

The integration of the sorbent processing and formulation, sorption-desorption application, metal recovery for reuse and the disposal of the spent biomass (Figure 9-1) is key to the successful industrial application of the technology developed.

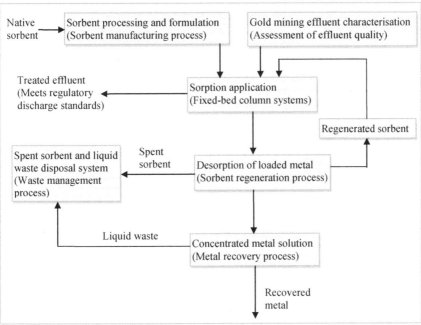

Figure 9-1: Generalised integrated single stage sorption systems for gold mining effluent treatment using locally available low-cost materials

9.6   Concluding remarks

This PhD research employed a sorption technique to remove heavy metals from gold mining effluent using natural and plant materials for sustainable treatment. An assessment of the effluent quality of a gold mining company in Ghana indicated that arsenic, copper and cyanide were the major pollutants in the process effluent. Arsenic and copper were successfully removed from the effluent by the studied materials. The research showed that the down-flow fixed-bed treatment configuration is an ideal system for the simultaneous removal of copper and arsenic from low concentration gold mining effluent, in addition to other heavy metals present in very low concentrations. The successful application of the sorption process developed in this study has a positive implication for industrial effluent treatment in a sustainable and cost effective manner.

9.7 References

Chandra Sekhar, K., Kamala, C.T., Chary, N.S. and Anjaneyulu, Y., 2003. Removal of heavy metals using a plant biomass with reference to environmental control. International Journal of Mineral Processing, 68(1-4): 37-45.

Febrianto, J., Kosasih, A.N., Sunarso, J., Ju, Y.-H., Indraswati, N., Ismadji, S., 2009. Equilibrium and kinetic studies in adsorption of heavy metals using biosorbent: A summary of recent studies. Journal of Hazardous Materials 162, 616-645.

Gadd, G.M., 2009. Biosorption: critical review of scientific rationale, environmental importance and significance for pollution treatment. Journal of Chemical Technology & Biotechnology 84, 13-28.

Ghana EPA, 2010. Environmental performance rating and disclosure: report on the performance of mining and manufacturing companies. Environmental Protection Agency, Accra, Ghana.

Gupta, V.K., Rastogi, A., 2008. Sorption and desorption studies of chromium(VI) from nonviable cyanobacterium Nostoc muscorum biomass. Journal of Hazardous Materials 154, 347-354.

Han, R. et al., 2006a. Biosorption of copper(II) and lead(II) from aqueous solution by chaff in a fixed-bed column. Journal of Hazardous Materials, 133(1-3): 262-268.

Kumar, R., Bhatia, D., Singh, R., Rani, S. and Bishnoi, N.R., 2011. Sorption of heavy metals from electroplating effluent using immobilized biomass Trichoderma viride in a continuous packed-bed column. International Biodeterioration & Biodegradation 65, 1133-1139.

Luo, X., Deng, Z., Lin, X. and Zhang, C., 2011a. Fixed-bed column study for Cu2+ removal from solution using expanding rice husk. Journal of Hazardous Materials, 187(1-3): 182-189.

Park, D., Yun, Y.-S. and Park, J., 2010. The past, present, and future trends of biosorption. Biotechnology and Bioprocess Engineering 15, 86-102.

Rahman, M.S., Islam, M.R., 2009. Effects of pH on isotherms modelling for Cu(II) ions adsorption using maple wood sawdust. Chemical Engineering Journal 149, 273-280.

Sarioglu, M., Güler, U.A., Beyazit, N., 2009. Removal of copper from aqueous solutions using biosolids. Desalination 239, 167-174.

Singh, S., Srivastava, V.C. and Mall, I.D., 2009. Fixed-bed study for adsorptive removal of furfural by activated carbon. Colloids and Surfaces A: Physicochemical and Engineering Aspects, 332(1): 50-56.

Singh, P., Bajpai, J., Bajpai, A.K. and Shrivastava, R.B., 2011. Fixed-Bed Studies on Removal of Arsenic from Simulated Aqueous Solutions Using Chitosan Nanoparticles. Bioremediation Journal 15, 148-156.

Vilar, V.J.P., Loureiro, J.M., Botelho, C.M.S. and Boaventura, R.A.R., 2008. Continuous biosorption of Pb/Cu and Pb/Cd in fixed-bed column using algae Gelidium and granulated agar extraction algal waste. Journal of Hazardous Materials, 154(1–3): 1173-1182.

Volesky, B., 2003. Sorption and Biosorption. BV Sorbex, Inc.,Montreal, Canada, pp. 34.

Samenvatting

Urbanisatie en industrialisatie in ontwikkelingslanden heeft geleid tot een groot probleem van het beheer van zowel vast als vloeibaar afval en Ghana is hierop geen uitzondering. Onbehandeld afvalwater van industrieën en huizen komt uiteindelijk terecht in rivieren en andere watersystemen die belangrijk zijn voor mens, fauna en flora. De meeste van deze rivieren worden gebruikt als bron voor drinkwater door plattelandsbewoners zonder enige vorm van behandeling met een groot risico voor hun gezondheid. In Ghana verontreinigt het onbehandelde afvalwater van de goudwinning de aquatische ecosystemen met zware metalen zoals koper (Cu), wat een bedreiging vormt voor het ecosysteem en de gezondheid van de mens. De ongewenste gevolgen van deze verontreinigende stoffen kunnen worden vermeden door behandeling van het mijnbouw afvalwater. Hoofdstuk 2 geeft een overzicht van relevante (bio)technologische mogelijkheden om zware metalen (zoals koper, arseen, lood en zink) uit afvalwater te verwijderen, met nadruk op afvalwater van de goudwinning en het gebruik van goedkope materialen als sorptiemiddel. Verschillende biologische en fysisch-chemische behandelingsprocessen worden besproken en vergeleken op basis van kosten, energiebehoefte, verwijderingsefficiëntie, beperkingen en voordelen. Sorptie met behulp van natuurlijke plantaardige materialen en met industrieel en agrarisch afval is een goed alternatief ter vervanging van conventionele methoden voor het verwijderen van zware metalen vanwege de kosteneffectiviteit, de efficiëntie en de lokale beschikbaarheid van deze materialen. Biosorptie zou meer bekendheid moeten krijgen als een opkomende technologie, effectief in het verwijderen van verontreinigingen in lage concentraties. De parameters die van invloed zijn op de (bio)sorptie, zoals ion concentratie, pH, sorptiemiddel dosering, deeltjesgrootte en temperatuur worden besproken. In het algemeen zijn technische toepasbaarheid, kosteneffectiviteit en complexiteit van het proces de belangrijkste factoren bij de keuze van de meest geschikte behandelingsmethode.

Een analyse van het afvalwater is nodig om te kunnen besluiten welke verontreinigingen verwijderd moeten worden. In hoofdstuk 3 wordt de kwaliteit besproken van het proces afvalwater van AngloGold Ashanti Limited, een bedrijf verantwoordelijk voor goudwinning in Ghana. De studie toonde aan dat het proces afvalwater van de goudwinning grote hoeveelheden gesuspendeerde vaste stof bevat, en daarom zeer troebel is. Arseen, koper en cyanide werden geïdentificeerd als de belangrijkste verontreinigende stoffen in het effluent met een gemiddelde concentratie van 10.0, 3.1 en 21.6 mg L^{-1}, respectievelijk. De concentraties arseen, koper, ijzer en cyanide (CN^-) overschreden de normen gesteld door de EPA van Ghana en daarom zal het afvalwater moeten worden behandeld voor lozing op het oppervlaktewater. In belangrijke mate wordt de samenstelling van het afvalwater bepaald door het gebruikte goudextratieproces en door de aard van het gouderts. Het afvalwater wordt geloosd op de Sansu dam, waar 99,9% van de TDS bezinkt en 99,7% van de turbiditeit verdwijnt, maar de concentraties koper, arseen en cyanide waren nog steeds hoog. Het geproduceerde effluent kan worden geclassificeerd als anorganisch met een hoog gehalte aan biologisch niet afbreekbare stoffen. De technologie voor de zuivering van dit afvalwater zou primair gericht moeten zijn op de verwijdering van koper en arseen. Sorptie met behulp van landbouwafval of plantaardig materiaal zou een goedkope optie kunnen zijn voor de behandeling van dit afvalwater.

Sorptie is een veelbelovende technologie om koper-rijk afvalwater uit de goudmijnen te behandelen. In hoofdstuk 4 staan de resultaten van de sorptie-eigenschappen voor Cu(II) van agrarische materialen, zoals kokosnootschil en -omhulsel, zaagsel en zaden van *Moringa oleifera*. Het Freundlich isotherm model beschreef de Cu(II) sorptie door kokosnootomhulsel (R^2 = 0.999) en zaagsel (R^2 = 0.993) zeer goed, de verwijdering van Cu(II) door *Moringa oleifera* zaden (R^2 = 0.960) goed en kokosnootschil redelijk (R^2 = 0.932). Kokosnootschil nam Cu(II) maximaal op (53.9 mg g^{-1}), volgens pseudo tweede-orde kinetiek (R^2 = 0.997). FTIR spectroscopie liet de aanwezigheid van functionele groepen in de biosorbents zien, waarvan sommige bij het sorptie-proces betrokken waren. SEM-EDX analyse bevestigde een uitwisseling van Mg(II) en K(I) voor Cu(II) op *Moringa oleifera* zaden en K(I) voor Cu(II) op de kokosschil. Deze studie toont aan dat kokosschil een belangrijk goedkoop biosorbent voor Cu(II) verwijdering uit anorganische afvalwater kan zijn. De resultaten laten zien dat ionenuitwisseling, neerslag en elektrostatische krachten bij de verwijdering van Cu(II) door de onderzochte biosorbents betrokken waren.

Het sorptiemiddel kan worden hergebruikt, maar moet dan worden geregenereerd met het juiste desorbent. In hoofdstuk 5 worden de sorptie/desorptie kenmerken van Cu(II) met betrekking tot kokosnootschil (CS) en met ijzeroxide bekleed zand (IOCS) beschreven. In batch testen werd gevonden dat CS een Cu(II) opname capaciteit heeft van 0,46 mg g^{-1} en een verwijdering van 93%, terwijl de opnamecapaciteit en verwijdering van IOCS voor Cu(II), respectevelijk 0.49 mg g^{-1} en 98% is. Desorptie-experimenten lieten zien dat HCl (0,05M) een efficiënt middel is voor het herstel van CS, met een gemiddelde desorptie efficiëntie van 96% (zelfs na acht sorptie en desorptie cycli). HCl (0,05M) heeft geen invloed op het vermogen van CS om koper op te nemen, zelfs na 8 cycli van sorptie en desorptie, maar verstoorde de ijzeroxide structuur van IOCS volledig binnen zes cycli. Deze studie toonde aan dat CS en IOCS beide goede sorptiemiddelen voor Cu(II) zijn, maar dat cyclische sorptie/desorptie met behulp van 0,05 M HCl alleen haalbaar is met CS.

Kinetische studies met betrekking tot adsorptie zijn van groot belang om een idee te krijgen van de capaciteiten van een absorbens en om inzicht te krijgen in het mechanisme. De sorptiekinetiek van Cu(II) voor kokosnootschil en *Moringa oleifera* zaden werd onderzocht met behulp van batch incubations in hoofdstuk 6. Om het mechanisme van het biosorptie-proces en de mogelijke proces-limiterende stappen te kunnen begrijpen werden kinetische modellen gebruikt voor de experimentele gegevens. De kinetische data worden het best beschreven door het pseudo second-order model met een correlatiecoëfficiënt (R^2) van 0.9974 en 0.9958 voor de kokosnootschil en zaden van de *Moringa oleifera*, respectievelijk. De koper opnamesnelheden voor kokosnootschil en *Moringa oleifera* zaden waren 9,6395 × 10^{-3} en 8,3292 × 10^{-2} mg.g^{-1}.min^{-1}, respectievelijk. De massaoverdracht coëfficiënten voor kokosnootschil (β_l = 1,2106 × 10^{-3} cm.s^{-1}) en *Moringa oleifera* zaden (β_l = 8,965 × 10^{-4} cm.s^{-1}) geven aan dat voor beide materialen het transport van Cu(II) uit de bulkvloeistof naar de vaste fase vrij snel plaats vindt. De resultaten toonden aan dat diffusie binnen de deeltjes de sorptiesnelheid bepaalt; diffusie vanuit de film kan echter niet worden verwaarloosd, vooral in het beginstadium van sorptie.

In hoofdstuk 7 wordt het functioneren van een fixed-bed kolom, gevuld met kokosnootschil, voor de biosorptie van Cu(II) ionen geëvalueerd met als parameters de kolom doorbraaktijd bij verschillende debieten, de bed-diepten en de Cu(II) concentraties. De bed diepte service tijd (BDST) en de Yoon-Nelson, Thomas en Clark modellen werden gebruikt voor het evalueren van de karakteristieke ontwerpparameters van de kolom. De Cu(II) biosorptie kolom functioneerde optimaal bij 10 mg.L^{-1} influent Cu(II) concentratie, 10 ml.min^{-1} debiet en 20 cm diepte van het bed. Onder deze optimale omstandigheden waren de doorbraaktijd en Cu(II) concentratie respectievelijk 58 h en 0,8 mg L^{-1}, waarna de koperconcentratie in het effluent de 1 mg.L^{-1} lozingsnorm (EPA, Ghana) overschreed. De opname van Cu(II) bedroeg 7,25 mg.g^{-1}, en dat is 14,5 keer hoger dan de waarde gevonden in de studie met hetzelfde materiaal voor de dezelfde beginconcentratie van Cu(II) (10 mg.L^{-1}). Het BDST model voorspelde de experimentele gegevens het beste in de 10% en 50% regio's van de doorbraakcurve. Het Yoon-Nelson model voorspelde de tijd die nodig is voor 50% doorbraak (τ) goed in alle onderzochte gevallen. De simulatie van de doorbraakcurve gaf een goede fit met het Yoon-Nelson model, maar de doorbraakcurve werd het best voorspeld door het Clark model. Het ontwerp voor een fixed-bed kolom voor Cu(II) verwijdering uit afvalwater door biosorptie met kokosnootschil kan worden gemaakt op basis van deze modellen.

Voor industriële toepassingen werd de verwijdering van zware metalen uit echt afvalwater van de goudmijn van AngloGold Ashanti (Obuasi, Ghana) bestudeerd van december 2011 tot juni 2012 in continue fixed-bed kolommen met kokosnootschil en met ijzeroxide bekleed zand met een constante doorstroomsnelheid van 150 mL.min^{-1}. Hoewel de zuiveringsinstallatie was bedoeld voor de verwijdering van koper en arseen, werden ook andere zware metalen (ijzer, lood en zink), aanwezig in zeer lage concentraties in het goudmijn afvalwater, verwijderd. In alle gevallen was de verwijderingsefficiëntie meer dan 98%. In één geval (run I) werd een totaal van 14.8 m^3 goudmijn afvalwater behandeld in 1608 h tot het moment dat de doorbraak van arseen plaats vond. Op dit punt waren koper, ijzer, lood en zink volledig verwijderd zonder sporen van de metalen in het behandelde afvalwater. De opname van koper bedroeg 16.11 mg.g^{-1}en dat is 2.23 keer zo hoog als de waarde verkregen in een laboratorium kolom getest met een enkelvoudig ion. Arseen en ijzer opnames bedroegen 12.68 en 5.46 mg.g^{-1}, respectievelijk. De studie toonde aan dat de down-stream fixed-bed behandeling van goudmijn afvalwater met lage concentraties een ideaal systeem is voor de gelijktijdige verwijdering van koper en arseen, en alsook de verwijdering van andere zware metalen in erg lage concentraties.

List of symbols

C_0 initial concentration of solute (mg L^{-1})
$C_{0, old}$ old initial concentration of solute (mg L^{-1})
$C_{0, new}$ new initial concentration of solute (mg L^{-1})
C_t outlet concentration at time t (mg L^{-1})
t   service time of the column (h)
Q   flow rate (mL min^{-1})
Q_{old} old flow rate (mL min^{-1})
Q_{new} new flow rate (mL min^{-1})
q_0 sorption capacity (mg g^{-1})
Z   bed depth of the column (cm)
C_B desired concentration of solute at breakthrough (mg L^{-1})
K_a sorption rate constant for the Bed Depth Service Time (BDST) model (L mg^{-1} h^{-1})
N_0 sorption capacity for the BDST model (mg L^{-1})
F   linear flow velocity of the metal solution through the bed (cm h^{-1})
M   dry mass of the biosorbent (g)
k_{TH} Thomas rate constant (mL min^{-1} mg^{-1})
k_{YN} Yoon-Nelson rate velocity constant (min^{-1})
τ time required for 50% adsorbate breakthrough (min) according to Yoon-Nelson
A   Clark model parameter
r   Clark model parameter
Y_i Solute concentration of component i in the solid phase (mg L^{-1})
k   Boltzmann constant (J K^{-1})
q^0_i Phase concentration of a single adsorbed component in equilibrium with C^0_i (mg L^{-1})
a,b Langmuir isotherm parameters for single component system (−)
a_i, b_i Langmuir isotherm parameters for multi-component system (−)

Greek letters

$\propto$ Elovich's initial sorption rate (mg g^{-1} min^{-1})
β_l mass transfer coefficient (cm s^{-1})
ρ_p density of the sorbent particles (g cm^{-3})
ε_p porosity of the adsorbent particles

List of publications

Peer-reviewed Journal Papers

1. **Acheampong, M.A.,** Lens, P.N.L., 2013. Biosorption applications for environmental technological processes using agricultural biomaterials. Trends in Environmental Biotechnology (In preparation).

2. **Acheampong, M.A.,** Lens, P.N.L., 2013. Treatment of gold mining effluent in pilot fixed bed sorption systems. Hydrometallurgy (Accepted).

3. **Acheampong, M.A.,** Ansa, E.D.O., Awuah, E., 2013. Biosorption of Pb(II) onto *Cocos nucifera* shell and *Moringa Oleifera* seeds. Chemical Engineering Communications (Accepted).

4. Ansa, E.D.O., **Acheampong, M.A.,** Nkrumah, F.K., Boakye-Yiadom, A.S., 2013. Escherichia coli attachment and detachment in algal and kaolin suspensions. Ghana Science Journal (In press).

5. **Acheampong, M.A.,** Dapcic, A.D., Yeh, D., Lens, P.N.L., 2013. Cyclic sorption and desorption of Cu(II) onto coconut shell and iron oxide coated sand. Separation Science and Technology (DOI:10.1080/01496395.2013.809362).

6. Pakshirajan, K., Worku, A., **Acheampong, M.A,** Lubberding, H., Lens, P.L., 2013. Cr(III) and Cr(VI) Removal from Aqueous Solutions by Cheaply Available Fruit Waste and Algal Biomass. Applied Biochemistry and Biotechnology 170, 498-513.

7. **Acheampong, M.A.,** Pakshirajan, K., Annachhatre, A.P., Lens, P.N.L., 2013. Removal of Cu(II) by biosorption onto coconut shell in fixed-bed column systems. Journal of Industrial and Engineering Chemistry 19, 841-848.

8. **Acheampong, M.A.,** Paksirajan, K., Lens, P.N.L., 2013. Assessment of the effluent quality from a gold mining industry in Ghana. Environmental Science and Pollution Research 20, 3799-3811.

9. **Acheampong, M.A.,** Pereira, J.P.C., Meulepas, R.J.W., Lens, P.N.L., 2011. Kinetics modelling of Cu(II) biosorption on to coconut shell and Moringa oleifera seeds from tropical regions. Environmental Technology 33, 409-417.

10. **Acheampong, M.A.,** Pereira, J.P.C., Meulepas, R.J.W., Lens, P.N.L., 2011. Biosorption of Cu(II) onto agricultural materials from tropical regions. Journal of Chemical Technology & Biotechnology 86, 1184-1194.

11. **Acheampong, M.A.,** Meulepas, R.J., Lens, P.N., 2010. Removal of heavy metals and cyanide from gold mine wastewater. Journal of Chemical Technology & Biotechnology 85, 590-613.

Conference Proceedings

1. Dapcic, A.D., **Acheampong, M.A.,** Lens, Piet N.L., 2012. Removal of Cu(II) from Aqueous Solution Using Coconut Shell and Iron Oxide Sand: Sorption and Desorption Cycle Studies. In: ***Proceedings of the 2012 University of South Florida Undergraduate Research Colloquium,*** Florida, USA (18 April, 2012).

2. **Acheampong, M.A.,** Lens, P.N.L., 2012. Effect of operating parameters on Cu(II) biosorption onto coconut shell in a fixed-bed column. In: ***Proceedings of the 2013 International Conference on Pollution and Treatment Technology,*** Hainan Island, China (2 - 4 January 2012).

3. **Acheampong, M.A.,** Lens, P.N.L., The Role of Biosorption in the Development of Plant Biotechnology. BIT's 1st Annual World Congress of Agricultural Biotechnology (WCAB-2011), Changchun, China (28-30 October 2011).

4. **Acheampong, M.A.,** Meulepas, R.J.W., Lens, P.N.L., 2011. Characterisation of the Process Effluent of AngloGold-Ashanti Gold Mining Company in Ghana. In: ***the proceedings of the 12th International Conference on Environmental Science and Technology,*** Rhodes Island, Dodecanese, Greece (8 -10 September 2011).

5. **Acheampong, M.A.,** Pereira, J.P.C., Meulepas, R.J.W., Lens, P.N.L., 2011. Equilibrium Isotherm and Kinetics Studies of Cu(II) Biosorption from Gold Mine Wastewater by Agricultural Materials. In: ***Proceedings of the 5th European Conference on Bioremediation,*** Chania, Greece (4 -7 July 2011).

6. **Acheampong, M.A.,** Pereira, J.P.C., Meulepas, R.J.W., Lens, P.N.L., 2011. Biosorption of Cu(II) and As(III) from Gold Mine Wastewater using Agricultural Materials: Biosorbents Screening and Equilibrium Isotherm Studies. In: ***Proceedings of the IWA International Conference on Water & Industry,*** Valladolid, Spain (1 - 4 May 2011).

About the author

Mike Agbesi Acheampong was born on 21 October 1969 in Kumasi, Ghana. He is an Environmental Chemical Engineer by training and has blended this background with expertise in Environmental Biotechnology. He obtained his B.Sc. (Chemical Engineering) and M.Sc. (Water Supply and Environmental Sanitation) degrees from the Kwame Nkrumah University of Science and Technology (KNUST), Kumasi, Ghana in 1996 and 2005, respectively. He worked for Unilever (Ghana) limited in 1996, Nestlé (Ghana) limited from 1997-1998, Demax Investments (1998-2003) and Kumasi Polytechnic (2003-date). He was appointed the head of the Chemical Engineering department of the polytechnic from 2007 to 2009 and acted as the Dean of the School of Engineering in 2008.

Mike Agbesi Acheampong joined the UNESCO-IHE Institute for Water Education in the Netherlands in 2009 as a PhD Research Fellow (2009-2013). His research work focused on the removal of heavy metals from gold mining wastewater using agricultural/plant materials as sorbents. Mike is an expert on metallurgical wastewater treatment using sorption technology. His PhD research work was supervised by Professor dr. ir. Piet N.L. Lens (Professor of Environmental Biotechnology). He has co-authored 17 papers on sorption and biosorption since 2009, including 6 conference papers. In 2011, he was promoted as a senior lecturer in the Chemical Engineering department of the Kumasi Polytechnic. Mike is a part-time farmer and a women right advocate. He is married to Rita Delali Acheampong with two lovely children, Manuella N.A. Acheampong and Mike K. Acheampong Jnr.

SENSE

Netherlands Research School for the
Socio-Economic and Natural Sciences of the Environment

C E R T I F I C A T E

The Netherlands Research School for the
Socio-Economic and Natural Sciences of the Environment
(SENSE), declares that

Mike Agbesi Acheampong

born on 21 October 1969 in Kumasi, Ghana

has successfully fulfilled all requirements of the
Educational Programme of SENSE.

Delft, 18 October 2013

the Chairman of the SENSE board

Prof. dr. Rik Leemans

the SENSE Director of Education

Dr. Ad van Dommelen

The SENSE Research School has been accredited by the Royal Netherlands Academy of Arts and Sciences (KNAW)

K O N I N K L I J K E   N E D E R L A N D S E
A K A D E M I E   V A N   W E T E N S C H A P P E N

The SENSE Research School declares that Mr. Mike Agbesi Acheampong has successfully fulfilled all requirements of the Educational PhD Programme of SENSE with a work load of 49 ECTS, including the following activities:

SENSE PhD Courses
o Environmental Research in Context
o Research Context Activity: Co-organizing the 3rd International Congress on Biotechniques for Air Pollution Control (28-30 September 2009 in Delft, The Netherlands)

Advanced MSc Courses
o Sustainable Wastewater Treatment and Reuse
o Solid Waste Management

Management and Didactic Skills Training
o Supervision of two MSc students
o Co-organising workshop on Development of Curriculum for Competence-Based Training Bachelor of Technology in Chemical Engineering, Kumasi Polytechnic, Kumasi, Ghana

Oral Presentations
o *Effect of operating parameters on Cu(II) biosorption onto coconut shell in a fixed-bed column.* International Conference on Pollution and Treatment Technology, 2-4 January 2013, Hainan Island, China
o *The Role of Biosorption in the Development of Plant Biotechnology.* BIT's 1st Annual World Congress of Agricultural Biotechnology, 28-30 October 2011, Changchun, China
o *Characterisation of the Process Effluent of AngloGold-Ashanti Gold Mining Company in Ghana.* 12th International Conference on Environmental Science and Technology, 8-10 September 2011, Rhodes Island, Dodecanese, Greece
o *Equilibrium Isotherm and Kinetics Studies of Cu(II) Biosorption from Gold Mine Wastewater by Agricultural Materials.* 5th European Conference on Bioremediation, 4 -7 July 2011, Chania, Greece
o *Biosorption of Cu(II) and As(III) from Gold Mine Wastewater using Agricultural Materials: Biosorbents Screening and Equilibrium Isotherm Studies.* IWA International Conference on Water & Industry, 1-4 May, 2011, Valladolid, Spain

SENSE Coordinator PhD Education

Drs. Serge Stalpers

T - #0417 - 101024 - C28 - 240/170/10 - PB - 9781138001657 - Gloss Lamination